APPLICATIONS OF LIQUID SCINTILLATION COUNTING

APPLICATIONS OF LIQUID SCINTILLATION COUNTING

DONALD L. HORROCKS

Scientific Instruments Division
Beckman Instruments, Inc.
Fullerton, California

ACADEMIC PRESS   New York and London   1974

A Subsidiary of Harcourt Brace Jovanovich, Publishers

ACADEMIC PRESS, INC.
111 Fifth Avenue, New York, New York 10003

United Kingdom Edition published by
ACADEMIC PRESS, INC. (LONDON) LTD.
24/28 Oval Road, London NW1

Library of Congress Cataloging in Publication Data

Horrocks, Donald L
    Applications of liquid scintillation counting.

    Includes bibliographical references.
    1.   Liquid scintillators.   I.   Title.
[DNLM: 1.   Scintillation counters.   QC787.C6 H816a
1974]
QC787.S34H67   539.7'75   73-9425
ISBN 0–12–356240–6

PRINTED IN THE UNITED STATES OF AMERICA

Dedicated to
my wife Marge and daughters
Andrea and Cindy

CONTENTS

Chapter III. SCINTILLATOR SOLUTIONS

Chapter IV. LIQUID SCINTILLATION COUNTERS
AND MULTIPLIER PHOTOTUBES

Chapter V. PARTICLE COUNTING TECHNIQUES

Chapter VI. PREPARATION OF COUNTING SAMPLES

Chapter VII. OXIDATION TECHNIQUES

Chapter VIII. COUNTING VIALS

Chapter IX. BACKGROUND

Chapter X. QUENCH CORRECTION METHODS

Chapter XI. DUAL-LABELED COUNTING

Chapter XII. CHEMILUMINESCENCE AND BIOLUMINESCENCE

PREFACE

It has been 10 years since a reference book has been published dealing solely with the subject of liquid scintillation counting ("The Theory and Practice of Scintillation Counting," by J. B. Birks, Pergamon Press, Oxford, 1964). In the intervening years there have been several excellent books published which have been the proceedings from conferences and symposia on liquid scintillation counting. These books often did not cover the complete scope of liquid scintillation counting but rather presented the special interests of research-type papers which were presented at the conferences and symposia. It therefore seemed desirable that another book be written which would include the many advances in liquid scintillation counting which have developed during these intervening years.

This book starts with discussions of the theory of liquid scintillation counting processes (Chapter II), the choice of the components of the liquid scintillator solution (Chapter III), and the development of the liquid scintillation counter and the multiplier phototubes (Chapter IV). These discussions are followed by considerations of the detection and measurement by liquid scintillation techniques of different types of particles produced by radionuclides (Chapter V).

The techniques and problems of sample preparations (homogeneous and heterogeneous) are presented along with a special section on oxidation techniques (Chapters VI and VII). The importance and difference of several types of counting vials are detailed (Chapter VIII), and typical sources of background and problems of reducing the background are discussed (Chapter IX). The sources of quenching in counting samples and methods of monitoring and correction for variable quench within samples are considered in detail (Chapter X).

Several special applications of liquid scintillation techniques are presented: dual-labeled counting (Chapter XI), chemiluminescence and bioluminescence (Chapter XII), radioimmunoassay (Chapter XIII), Cerenkov counting (Chapter XIV), pulse shape discrimination (Chapter XV), flow cell counting (Chapter XVI), and large-volume counters (Chapter XVII). In conclusion, the statistical considerations involved in determining the reliability and accuracy of data obtained by nuclear counting techniques are considered (Chapter XVIII).

Of course it is impossible to cover all the applications of liquid scintillation counting in any one book. The use of liquid scintillation counting has become quite common in the biosciences, medicine, environmental and space sciences, chemistry, physics, and other fields, and it seems to be an impossible task for anyone to keep fully aware of all these applications. One of the more important recent applications which has been included in this book is the use of liquid scintillation counting for radioimmunoassay (RIA) testing. RIA has such a great potential in the in-vitro diagnostic determination of many nonnormalcies of the body functions that the use of liquid scintillation counting in this area may very well exceed its use in all other areas in a matter of a few years. For this reason and because of the extreme necessity for reliable and accurate answers, it seems to be a desirable goal that those doing RIA by liquid scintillation counting have an understanding of the limits, advantages, and trouble areas they will face. It is hoped that this book will fill these needs as well as serve as a reliable source of information for others already using or starting to use liquid scintillation counting techniques.

ACKNOWLEDGMENTS

I personally wish to express my indebtedness to all those whose published and unpublished results have contributed so much to this book and without which this book could not have been written. I am indebted to the staffs and managements of Argonne National Laboratory (Argonne, Illinois) and Beckman Instruments, Inc. (Fullerton, California) for their support and assistance during the writing and publication of this book. Most of the typing was performed by Ms. M. Volkert and E. Kelley of Beckman Instruments, Inc., for which I am deeply appreciative. Finally, without the moral support of my family it would have been impossible to write this book.

APPLICATIONS OF LIQUID
SCINTILLATION COUNTING

CHAPTER I

INTRODUCTION

Shortly after it was demonstrated that organic solutions could be induced to fluoresce upon excitation by nuclear radiations, the potential of liquid scintillation counting was realized by some very far-sighted persons. The rapid development of efficient scintillation solutions and the coincidence counter opened up the whole field of biological nuclear tracing. Prior to the liquid scintillation counter, ^{14}C had to be counted as a solid ($BaCO_3$) or gas (CO_2), with many problems associated with the measurement of the amounts of ^{14}C. It was almost impossible to assay ^3H except by gas counting.

The internal sample technique led to increased sensitivities for the measurement of all low-energy beta emitters, especially ^3H and ^{14}C. However, many new problems were encountered as investigators sought to lower the limits of detection and to introduce a wider variety of samples into the scintillator solution.

Modern liquid scintillation counters and advanced sample preparation techniques have increased the potential of liquid scintillation well beyond even the most optimistic concepts of the early workers in this field. Large amounts of water can now be counted directly as emulsions in liquid scintillation systems with sensitivities which allow for the detection of picocuries (10^{-12} Ci) of tritium in water.

Early History of the Liquid Scintillation Method

As early as 1937, Professor H. P. Kallmann, in Germany, noted that certain organic materials fluoresced under ultraviolet light (1). When in a darkroom he observed fluorescence from materials in the seams of clothing. His investigations were interrupted by World War II, at the end of which he was able to obtain enough naphthalene to make a large crystal, which he showed could be used to detect radiations such as alphas, betas, and gammas. Because of the need for large-volume detectors of varying sizes and shapes, in 1947 Kallmann showed that aromatic solvents with certain dissolved solutes were efficient scintillation sources when subjected to nuclear radiations (2).

In the period 1949–1950, three papers appeared in the literature which clearly demonstrated that organic solutions could be used as scintillation counters. In 1950 Kallmann used toluene and xylene solutions of fluorene, carbazole, phenanthrene, and anthracene as scintillation detectors (3). Ageno and co-workers in Rome counted alpha, beta, and gamma radiations in xylene solutions of naphthalene (4), and Reynolds and co-workers at Princeton demonstrated the coincidence method of a counter with benzene and xylene solutions of *p*-terphenyl, noting that the efficiency was as high as naphthalene crystals (5). Other early works were reported by Hofstadter and co-workers (6) and Belcher (7).

First Liquid Scintillation Counters

Most of the early liquid scintillation counters were based on measurement of the anode current from a single multiplier phototube (MPT) as a function of the level of radioactivity. In 1950 Reynolds and co-workers published the results of a coincidence liquid scintillation counter they had developed which used a scaler to record the level of radioactivity (5). Figure I-1 shows the diagram of this simple but far-sighted design.

In 1952 both Belcher in London (7) and Hayes and co-workers (8) in Los Alamos, New Mexico, published the development of a single MPT counter that measured individual pulses (counts) and was able to discriminate the pulse amplitude as directly related to the energy of the particle that excited the organic scintillator. The Belcher and Los Alamos counters

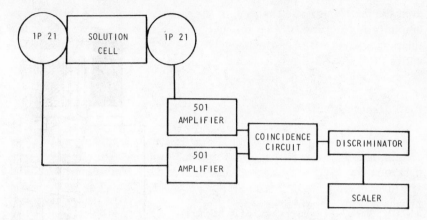

Fig. I-1. Coincidence counter described by Reynolds and co-workers (5).

are shown diagrammatically in Figs. I-2 and I-3, respectively. The single MPT counters were not suitable for the counting of low-energy beta emitters (i.e., ^3H), because the spontaneous production of counts by the MPT (noise or background) was very high.

The prototype of the early commercial coincidence liquid scintillation counter was developed at the Los Alamos Scientific Laboratory (U.S. Atomic Energy Commission) by Hiebert and Watts (9) (see Fig. I-4). About the same time, and independently, Arnold (10) at the University of Chicago also developed a similar design for a coincidence-type liquid scintillation counter.

Packard Instrument Company of Downers Grove, Illinois, under the direction of its founder, Lyle Packard, saw the great potential of liquid scintillation counting and undertook to develop the first commercial liquid scintillation counter after the design of Hiebert and Watts. The first counter was subsequently installed at the Argonne Cancer Research Hospital, University of Chicago, for Dr. LeRoy and co-workers (11).

The development of liquid scintillation counters over the years with relation to the counting efficiencies for tritium and carbon 14 and the background is shown in Table I-1. The improvement has been steady, with innovations in instrument design and the development of new photo-tubes contributing to the progress.

Early Scintillator Solutes

The earliest organic scintillators were used in the form of crystals. As a logical extension of the work with the crystals, the early liquid scintillators

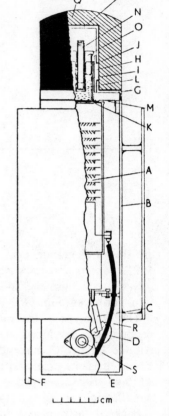

Fig. I-2.   Single MPT counter described by Belcher (7).

Table I-1

Development of Liquid Scintillation Counter Performance

	For ³H		For ¹⁴C	
Year	Counting efficiency (%)	Background (cpm)	Counting efficiency (%)	Background[a] (cpm)
1954	10	80	75	60
1962	25	55	80	30
1964	40	30	85	25
1969	60	20	90	16
1972	65	18	97	11

[a] Background above ³H endpoint. Total background is the sum of ³H background and last column.

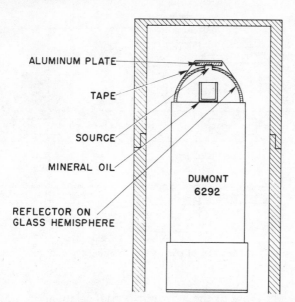

ALUMINUM PLATE

TAPE

SOURCE

MINERAL OIL

DUMONT
6292

REFLECTOR ON
GLASS HEMISPHERE

Fig. I-3. Liquid scintillation counter used by Hayes and co-workers (8) to evaluate scintillator solutes.

employed solutions of the crystalline materials. Some of the compounds are listed in Table I-2.

The very excellent and timely work of Hayes and co-workers at Los Alamos was the most fortunate piece of research on scintillator solutes (12–15). They evaluated literally hundreds of organic compounds for their use in liquid scintillation counting, most of which, amazingly, are still the best scintillator solutes 20 years later.

Table I-3 lists just a small portion of the compounds that they rated in the oxazole and oxadiazole series, of which PPO is still the most widely used scintillator solute.

Other Early Contributions

There are so many people who contributed to the development of liquid scintillation counting that it would be impossible to mention all of them in this book. Lack of space necessitates the omission of some who deserve recognition, for which we apologize. However, some of the early workers who stand out in this author's opinion are cited here.

Swank, Buck, and co-workers at Argonne National Laboratory (U.S. Atomic Energy Commission) measured the decay times of organic scintillators in solution (16) and showed that the decay constants were many

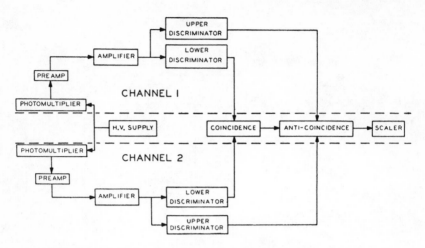

Fig. I-4.   Block diagram of Los Alamos counting apparatus, 1952 (9).

times faster than those of the inorganic crystal scintillators, i.e., NaI(Tl), with $\tau = 230$ nsec. They also observed that impurities (especially oxygen) altered the decay time, usually shortening it. The decay times published by Swank and Buck in 1955 are listed in Table I-4. Of course, many of these values have since been remeasured with a greater accuracy.

Of course, the work of Förster, from 1949 to the present, on the theory and mechanism of energy transfer processes in liquids, has been invaluable (17–22). The Förster resonance energy transfer theory explained the quantitative energy transfer from excited solvent molecules to the solute molecules, even when the concentration of the solute molecules was too low, as required by collision theory. The energy transfer theory has also been investigated by Galanin (23–29). This theory is even more important in explaining the quantitative transfer of excitation energy from primary solutes to secondary solutes, which are present at only a fraction of the primary solute concentration.

Birks and co-workers at the University of Manchester, England, have contributed greatly to the understanding of the relationship between photophysical processes and scintillation processes (30–35). These workers showed that the same excited state is responsible for the observed fluorescence when the excitations are produced directly in the solute molecule by absorption of uv radiation as when they are produced by ionizing particle excitations in the solvent molecules (36).

Kallmann and co-workers at New York University made early contributions to the study of energy transfer process (37–49), by adding to the understanding of the mechanism of quenching. The relationship between quencher concentration and decreased fluorescence intensity led to the often used Kallmann parameters.

Of course, many investigators contributed to the knowledge of sample preparations. The advantage of the internal sample method depends on either dissolving the sample in the scintillator solution or suspending it in a finely divided state. To count tritium with reasonable efficiencies, it is necessary that the sample be in intimate contact with the scintillation medium. The ideal sample is one that is completely soluble in the scintillator solution and which does not inhibit the process of producing the photons emitted.

Table I-2

Some organic compounds which early investigators found would scintillate as single crystals and when dissolved in aromatic hydrocarbon solvents

Compound	Structure
Naphthalene	
Anthracene	
p-Terphenyl	
trans-Stilbene	
Phenanthrene	
Carbazole	

Table I-3

Relative scintillation efficiency of several compounds used as solutes in toluene

Compound	Common name or abbreviation	Relative scintillation efficiency[a]
	p-Terphenyl	94
	PPO	100
	—	22
	PPD	87
	—	<0.2
	PBO	114
	PBO	114
	PBD	120

[a] Relative to 100 for PPO at 3 g/liter in toluene.

Table I-4

Scintillation data reported by Swank and Buck[a] in 1955

Solute (g/liter)	Solvent	RPH[b]	Decay constant (10^{-9} sec)	Present accepted[c] decay constant (10^{-9} sec)
p-Terphenyl, 5.0	Toluene	0.35	2.2	1.0
PPO, 3.0	Toluene	0.40	<3.0	1.4
PBD, 8.0	Toluene	0.49	<2.8	1.2

[a] R. K. Swank and W. L. Buck, *Rev. Sci. Instrum.* **26**, 15 (1955).

[b] RPH = relative pulse height. Relative to the response per unit energy excitation for anthracene crystal equal to 1.00.

[c] J. B. Birks, "Photophysics of Aromatic Molecules." Wiley (Interscience), New York, 1970.

Previous Books

Many fine books have been published on the subject of liquid scintillation counting, ranging from proceedings of conferences and symposia to monographs and reference books. It is hoped that this book will be a complement to these previous publications, a partial list of which follows.

Birks, J. B., "Scintillation Counters." Pergamon, Oxford, 1953.

Curran, S. C., "Luminescence and the Scintillation Counter." Butterworth, London, 1953.

Bell, C. G., and Hayes, F. N. eds., "Liquid Scintillation Counting." Pergamon, Oxford, 1958.

Daub, G. H., Hayes, F. N., and Sullivan, E. eds., *Proc. Conf. Organic Scintillation Detectors, Univ. of New Mexico, 1960,* TID-7612. U. S. At. Energy Commission, Washington, D. C., 1961.

"Tritium in the Physical and Biological Sciences," Vols. 1 and 2. IAEC, Vienna, 1962.

Schram, E., "Organic Scintillation Detectors." Elsevier, Amsterdam, 1963.

Birks, J. B., "The Theory and Practice of Scintillation Counting." Pergamon, Oxford, 1964.

Rapkin, E., Liquid scintillation counting 1957–1963: A review. *Int. J. Appl. Radiat. Isotop.* **15,** 69 (1964).

Rothchild, S. ed., *Advan. Tracer Methodol.* **1** (1962); **2** (1965); **3** (1966).

Berlman, I. B., "Handbook of Fluorescence Spectra of Aromatic Molecules." Academic Press, New York, 1965.

Horrocks, D. L., ed., "Organic Scintillators." Gordon & Breach, New York, 1968.

Parmentier, J. H., and TenHaaf, F. E. L., Developments in liquid scintillation counting since 1963. *Int. J. Appl. Radiat. Isotop.* **20,** 305 (1969).

Bransome, Jr., E. D., ed., "The Current Status of Liquid Scintillation Counting." Grune & Stratton, New York, 1970.

Horrocks, D. L., and Peng, C. T., eds., "Organic Scintillators and Liquid Scintillation Counting." Academic Press, New York, 1971.

Dyer, A., ed., "Liquid Scintillation Counting," Vol. 1. Heyden, London, 1972.

There are also a large number of very fine publications on liquid scintillations, which have been issued as technical publications of commercial companies for the manufacture of liquid scintillation counters and chemicals. Some of these companies are as follows:

Arapahoe Chemicals, Boulder, Colorado
Beckman Instruments, Inc., Fullerton, California
Intertechnique Instruments, Inc., Dover, New Jersey and Plaisir, France
Koch-light Laboratories, Ltd., Colnbrook, England
New England Nuclear Corp., Boston, Massachusetts
Nuclear-Chicago Corp., Des Plaines, Illinois
Nuclear Enterprises, Ltd., Edinburgh, Scotland
Philips, Eindhover, Netherlands
Packard Instruments Company, Downers Grove, Illinois.

References

1. A Young physicist at seventy: Hartmut Kallmann, *Phys. Today* **19**, 51–54 (1966).
2. H. Kallmann, *Z. Naturforsch. A* **2**, 439, 262 (1947).
3. H. Kallmann, *Phys. Rev.* **78**, 621 (1950).
4. M. Ageno, M. Chiozzotto, and R. Querzoli, *Phys. Rev.* **79**, 720 (1950).
5. G. T. Reynolds, F. B. Harrison, and G. Salvini, *Phys. Rev.* **78**, 488 (1950).
6. R. Hofstadter, S. H. Liebson, and J. O. Elliot, *Phys. Rev.* **78**, 81 (1950).
7. E. H. Blecher, *J. Sci. Instrum.* **30**, 286 (1953).
8. F. N. Hayes, R. D. Hiebert, and R. L. Schuch, *Science* **116**, 140 (1952).
9. R. D. Hiebert and R. J. Watts, *Nucleonics* **11** (12), 38 (1953).
10. J. R. Arnold, *Science* **119**, 155 (1954).
11. L. E. Packard, Private communication, 1953.
12. F. N. Hayes, B. S. Rogers, P. Sanders, R. L. Schuch, and D. L. Williams, Rep. LA-1639. Los Alamos Sci. Lab., Los Alamos, New Mexico, 1953.
13. F. N. Hayes, D. G. Ott, V. N. Kerr, and B. S. Rogers, *Nucleonics* **13** (12), 38 (1955).
14. F. N. Hayes, D. G. Ott, and V. N. Kerr, *Nucleonics* **14** (1), 42 (1956).
15. F. N. Hayes, *in* "Liquid Scintillation Counting" (C. G. Bell and F. N. Hayes, eds.), p. 83. Pergamon, Oxford, 1958.
16. R. K. Swank and W. L. Buck, *Rev. Sci. Instrum.* **26**, 15 (1955).
17. Th. Förster, *Ann Phys. (Leipzig)* **2**, 55 (1948).
18. Th. Förster, *Z. Naturforsch. A* **4**, 321 (1949).
19. Th. Förster, "Fluorescenz Organischer Verbindungen." Vandenhoeck & Ruprecht, Gottingen, 1951.
20. Th. Förster, *Discuss. Faraday Soc.* **27**, 7 (1959).
21. Th. Förster, *Z. Elektrochem.* **64**, 157 (1960).
22. Th. Förster, *Radiat. Res.* **2**, 326 (1960).
23. M. D. Galanin, *Zh. Eksp. Teor. Fiz.* **21**, 126 (1951).
24. M. D. Galanin and I. M. Frank, *Zh. Eksp. Teor. Fiz.* **21**, 114 (1951).
25. M. D. Galanin. *Sov. Phys. JETP* **1**, 317 (1955).
26. T. P. Belikova and M. D. Galanin, *Opt. Spectrosc.* (USSR) **1**, 168 (1956).
27. T. P. Belikova, M. D. Galanin, and Z. A. Chizhikova, *Bull. Acad. Sci. USSR Phys. Ser.* **20**, 349 (1956).
28. T. P. Belikova and M. D. Galanin, *Bull. Acad. Sci. USSR Phys. Ser.* **22**, 48 (1958).
29. M. D. Galanin, *Tr. Fiz. Inst. Akad. Nauk SSSR* **12**, 3 (1960).
30. J. B. Birks, *IRE Trans. Nucl. Sci.* **NS-7** (2, 3), 2 (1960).
31. J. B. Birks, *Proc. Conf. Organic Scintillation Dectectors; Univ. of New Mexico* (G. H. Daub, F. N. Hayes, and E. Sullivan, eds.), p. 12. TID-7612 U. S. At. Energy Commission, Washington, D.C. 1961.
32. J. B. Birks and K. N. Kuchela, *Proc. Phys. Soc. London* **77**, 1083 (1961).
33. J. B. Birks, *Nucl. Electron. Proc. Conf. Belgrade, 1964* **1**, 17. IAEA, Vienna, 1962.
34. J. B. Birks, "The Theory and Practice of Scintillation Counting." Pergamon, Oxford, 1964.
35. J. B. Birks, "Photophysics of Aromatic Molecules." Wiley (Interscience), New York, 1970.
36. J. B. Birks, C. L. Braga, and M. D. Lumb, *Brit. J. Appl. Phys.* **15**, 399 (1964).
37. H. Kallmann and M. Furst, *Phys. Rev.* **79**, 875 (1950).
38. H. Kallmann and M. Furst, *Phys. Rev.* **81**, 853 (1951).
39. H. Kallmann and M. Furst, *Nucleonics* **8** (3), 32 (1951).
40. M. Furst and H. Kallmann, *Phys. Rev.* **85**, 816 (1952).
41. M. Furst and H. Kallmann, *Phys. Rev.* **94**, 503 (1954).

42. M. Furst and H. Kallmann, *Phys. Rev.* **96,** 902 (1954).
43. M. Furst and H. Kallmann, *J. Chem. Phys.* **23,** 607 (1955).
44. M. Furst and H. Kallmann, *Phys. Rev.* **109,** 646 (1958).
45. H. Kallmann and M. Furst, *in* "Liquid Scintillation Counting" (C. G. Bell and F. N. Hayes, eds.) p. 3. Pergamon, Oxford, 1958.
46. F. H. Brown, M. Furst, and H. Kallmann, *J. Chim. Phys.* **55,** 688 (1958).
47. G. K. Oster and H. Kallmann, *Nature (London)* **194,** 1033 (1962).
48. G. K. Oster and H. Kallmann, *Int. Symp. Luminescence* (N. Riehl, H. Kallmann, and H. Vogel, eds.), p. 31. Thiemig, München, 1966.
49. G. K. Oster and H. Kallmann, *J. Chim. Phys.* **64,** 28 (1967).

CHAPTER II

BASIC PROCESSES

The detection of energy, in the form of nuclear emanations, by organic solutions is dependent on a charged particle (the nuclear emanations or the product of interactions of the nuclear emanations) producing a number of excited molecules in the organic solution. These excited molecules will either emit photons or efficiently transfer the energy to an acceptor (solute) which in turn will emit the photons. Each excited molecule can emit only one photon and the energy of that photon will be within a limited range. The number of photons emitted will depend on the number of excited molecules produced by the ionizing particle. The photons can be measured by collection on the face of a multiplier phototube which will convert them into an electrical pulse.

Interaction of Ionizing Radiation with Matter

Nuclear radiations will interact with molecules in many ways. The ionizing radiations (i.e., electrons, protons, alphas, etc.) will make a "track" as the particle passes through a material. Along the track there will be a large number of molecules which will be given energy from the particles, but the majority of the molecules in the material will not be affected at all; that is, the effects are very local. The energy loss by the ionizing particle will produce ions, excited molecules, free radicals, secondary particles (scattered or newly formed), and the energy of the molecules (vibrational, rotational, and kinetic). Along the track the concentration of the excited and ionized molecules will determine the types of processes (chemical and physical) that will occur as the result of the interactions which follow the passage of the particle.

The types of chemical reactions that occur are neutralization, free radical reactions, decomposition, excimer formation, and many others. Some of the physical processes that occur are X-ray emission, fluorescence, phosphorescence, energy migration, energy transfer, radiationless deactivation, and many others. In many systems, including liquid scintillation solutions, almost all of these processes will occur. The final result, the emission of photons, will be effected by the competition between all of these reactions.

The production of ions can occur by at least two main processes: (1) direct scattering (billiard-ball-type collisions) by orbital electrons in the atoms that make up the molecules of the material (Fig. II-1a), and (2) excited molecules which possess energy in excess of the ionization potential of the molecule (Fig. II-1b). Depending on which orbital electron is removed from the atom, there may also be X rays emitted during the scattering-type collisions.

Collision and scattering by electrons in the outer orbits usually lead to an ion with the formation of two electrons with discrete energies. The energy of the primary scattered electron will be decreased only by the binding energy of the orbital electron and the energy that goes into momentum conservation and imparted kinetics of the ejected electron (Fig. II-2a). If the ejected electron is from one of the inner orbital shells, a characteristic (or series of characteristic) X ray will also be produced (Fig. II-2b).

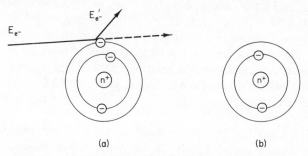

(a)　　　　　　　　　　　　(b)

Fig. II-1.　Schematic of direct scattering by orbital electrons.

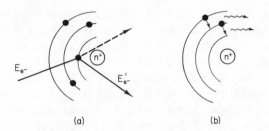

Fig. II-2.   X-Ray characteristic of the energy difference between the two shells involved in the photoelectric effect.

The same types of processes can occur with molecules except that the molecules can utilize some of the excess energy (in vibrations and rotations) and can remain intact in an excited state. Many molecules can spontaneously give up this excess energy upon returning to their normal ground state (no excess energy) by the emission of a photon (luminescence). The initial effect of the ionizing radiation will produce some or all of the following species, depending on the complexity of the molecules:

$$
\begin{aligned}
\text{ions:} \quad & RH^+,\ RH^-\ (A^+,\ A^-)\\
\text{radicals:} \quad & R\cdot,\ RH\cdot\ (A\cdot)\\
\text{excited molecules:} \quad & RH^*\ (A^*)\\
\text{fragments:} \quad & G,\ H,\ J,\ \text{etc. (or } G^+,\ H^+,\ J^+,\ \text{etc.).}
\end{aligned}
$$

These initial products can undergo further reactions:

$$
\begin{aligned}
A^+ + e^- &\longrightarrow A^* \quad\text{or}\quad A & \text{ion recombinations}\\
A^* + Q &\longrightarrow A + Q & \text{quenching processes}\\
A + G^+ &\longrightarrow AG^+ & \text{ion scavenging}\\
AG^+ + HJ &\longrightarrow AJ^+ + GH & \text{ion–molecule reactions}\\
A^* + A^* &\longrightarrow A^+ + A + e & \text{autoionization}\\
A^* + A &\longrightarrow A + A^* & \text{energy transfer.}
\end{aligned}
$$

In solids and liquids these secondary reactions occur in or very near the particle track where the concentration of the reactants is highest. The overall probability of these secondary reactions is directly related to the concentration of the reacting species. The concentration of these species is related to the specific ionization of the exciting particle. The specific ionization dE/dx is the energy loss by the particle per path length. The specific ionization of different types of ionizing particles as a function of the range (function of path length) is shown in Fig. II-3.

Table II-1 lists a few data on the specific ionization and range of electrons in a medium of 1-g/cm³ density. A 480-keV electron produces just over two primary ions per micron, while an electron which is 33 times slower

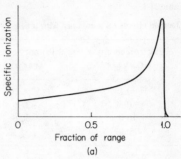

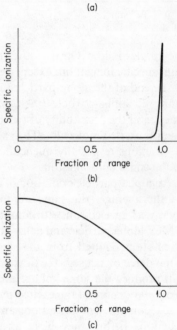

Fig. II-3. Relationship between specific ionization and fraction of range for (a) alpha, (b) beta, and (c) fission fragments.

(0.5 keV) produces ions at the rate of 425 ions/μm, or in 210 times greater concentration. The 470-keV electron loses about 207 eV of its energy in passing through 1 μm of material ($\rho = 1$ g/cm^3), while a 0.5-keV electron dissipates energy at such a rapid rate that it is stopped in a very small fraction of a micron of the same material. The very rapid increase in energy loss per micron with decreasing velocity (energy) is characteristic of all charged particles. This partly explains why delta (δ) rays and secondary electrons produce intense local concentrations of ionizations.

Electrons have a uniform, low specific ionization, except at the very end of the track. Therefore, the different species produced along the track are usually at a low concentration, and the probability of interactions between

Table II-1

Electron effects on a medium with a density of 1 g/cm³

Electron energy (keV)	Primary ions (per μ)	Energy dissipation (keV/μ)	Range (μ)[a]
0.1	1700.0	33.2	0.0030
0.5	425.0	18.6	0.0196
10.0	30.0	2.30	2.5
480.0	2.1	0.207	1650.0

[a] $\mu = 10^{-4}$ cm.

the species is low. On the other hand, alpha particles have a nonuniform, high specific ionization, except again there is a very high specific ionization at the end of the alpha-particle range. The concentration of various species along an alpha-particle track is much higher, compared to an electron track, and the probability of reactions is greater. Part of these reactions will involve excited molecules. The excited molecules that undergo certain types of reactions (especially quenching-type processes) will not be able to lead to the emission of photons.

An alpha-particle excitation of a liquid scintillation solution will produce, on the average, the same number of photons per megaelectron volt of energy as an electron with about one-tenth of its energy (1–5). That is, a 5-MeV alpha particle and a 0.5-MeV electron will produce the same number of photons emitted from the same liquid scintillation. Assuming that the same detector is used, the 5-MeV alpha particle and the 0.5-MeV electron will produce the same voltage amplitude pulse. This is due to the fact that nearly 90% more of the excited molecules produced along the alpha-particle track, compared to those along an electron track, are quenched by competing reactions before they have an opportunity to transfer their excitation energy to the fluors which will emit photons.

A proton has a specific ionization somewhat less than an alpha particle, but much greater than an electron. The electron equivalent light output of a proton is about one-fourth of its energy (6). Fission fragments from ^{252}Cf fission have about 180 MeV energy and an extremely short range (i.e., high specific ionization). The 180-MeV fission process produces light equivalent to an electron with about 2.4 MeV energy (7). Table II-2 summarizes the relative photon yields as a function of the type of particle.

In solutions the relationship between the components is very important. For example, consider a liquid scintillation solution which has water present (either as the sample or as a secondary solvent for the sample). When water is irradiated, peroxide (H_2O_2) is formed which may chemically react with other components in the system, leading to photon production through chemiluminescence. Samples with very high activity and energy may produce chemiluminescence in this manner.

Scintillation in Organic Material

A certain fraction of the ions and excited molecules produced in organic materials will lead to luminescence. In some cases the yield is very low or the energy is such that it is very difficult to detect the luminescence. However, there are many organic materials which are efficient converters of nuclear energy into light. The spectrum (energy of photons) of emitted light will be characteristic of the particular molecule and its excited electronic energy levels. Luminescence has been observed in a very wide variety of organic materials. A partial list includes gases, liquids (pure), solids (crystals, mixed crystals), plastics, and solutions.

Table II-2

Relative scintillation yield for different types of ionizing particles

Type of particle	Fraction of particle energy converted to photons compared to electrons
Electrons (> 80 keV)	1.00
Protons (1–10 MeV)	0.20–0.50
Alphas (4–6 MeV)	0.08–0.12
Fissions (180 MeV)	0.013

There are basic differences between the scintillation processes in organic materials and those in inorganic materials. The energy of the ionizing particle is partly transformed into luminescence emissions. The luminescence has a characteristic spectrum which seems to be a property associated with conjugated and aromatic organic molecules. It is an inherent molecular property and arises from the electronic structure of the molecule. Many of these organic molecules form molecular crystals in which the molecules are held together by weak van der Waals forces, and where they maintain much of their individual identity, electronic structure, and luminescence properties. Many organic molecules show the same luminescence spectrum dissolved in solution as in pure crystalline state (except in certain cases where dimers are formed).

Absorption Spectrum

Saturated hydrocarbons, i.e., cyclohexane, which contain no π electrons show no optical absorption at energies less than ~ 6 eV ($\gtrsim 200$ nm). However, many molecules which contain as part of their electronic structure, nonlocalized π electrons require much less energy to cause electronic excitations. Generally three (or more) π-electronic absorption bands are readily observed which correspond to transitions from the singlet ground state (S_0) into the singlet π-electronic excited states (S_1, S_2, S_3, etc.). A simplified diagram of the energy levels involved in a typical excitation of an organic molecule with π-electronic levels is shown in Fig. II-4.

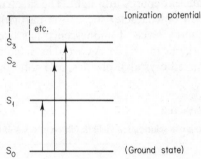

Fig. II-4. A simplified molecular energy level diagram showing the singlet ground state (S_0) and the excited singlet states.

The probability of a given excited state being formed by absorption of energy from an ionizing particle is proportional to the extinction coefficient ϵ for the optical absorption into that singlet level. Consider the aromatic solvents benzene, toluene, and *p*-xylene which are commonly used as solvents in liquid scintillators. Table II-3 lists the extinction coefficients for the transitions from the ground state to the S_1, S_2, and S_3 excited singlet states. The probability of producing these excited states from the passage of an electron through these solvents has also been measured. The S_3 state is predominantly produced (8–10).

However, the S_1 excited state is the state that is responsible for the fluorescence emission from all (with very few exceptions) excited organic molecules (11–13). The upper excited states are converted to the lower excited states by the process of internal conversion. Internal conversion is a radiationless process—i.e., no photons are emitted. For each excited state there is a competition between photon emission and internal conversion (and other radiationless processes). Most upper excited states ($> S_2$) undergo deexcitation by internal conversion. Only the S_1 state of many organic molecules has a higher probability for photon emission than for radiationless transition between the S_1 state and ground-state vibrational levels.

Table II-3

Fast electron excitation probabilities related to optical extinction coefficient

	Transition	Optical extinction coefficient	Fraction of transitions produced by fast electrons
Benzene	$S_0 \rightarrow S_1$	280	0.001
	$S_0 \rightarrow S_2$	8800	0.126
	$S_0 \rightarrow S_3$	68,000	0.873
Toluene	$S_0 \rightarrow S_1$	260	0.001
	$S_0 \rightarrow S_2$	7900	0.126
	$S_0 \rightarrow S_3$	55,000	0.873
p-Xylene	$S_0 \rightarrow S_1$	700	0.001
	$S_0 \rightarrow S_2$	8600	0.126
	$S_0 \rightarrow S_3$	59,000	0.873

Energy Transfer

Most aromatic solvents are not good scintillators by themselves. Therefore solutes that are efficient scintillators are added to the solvents. However, the energy has to find its way from the excited solvent molecules to the efficient fluor molecules. The efficiency of the transfer process is a function of the solute concentration (14–16). Figure II-5 shows typical plots for the scintillation yield as a function of the concentration of the solutes—PPO, *p*-terphenyl, and butyl-PBD.

The scintillation yield increases very rapidly with concentration of solute at low solute concentrations. The scintillation yield reaches a maximum and for some solutes does not change with increasing solute concentration (at least over limited solute concentration range).

Solvent–Solvent Transfer

The primary energy transfer process occurs between solvent molecules. Two theories are presently being debated regarding the actual energy transfer mechanism. One theory, proposed by Birks (17, 18), involves the formation of solvent excimers, which upon breaking apart find the excitation energy on the previously unexcited solvent molecule. This solvent excimer formation and breaking up occurs many times, allowing the energy to

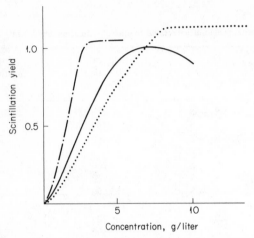

Fig. II-5.   Relative scintillation yield as function of the concentration of three typical solutes: PPO (—), butyl-PBD (......), and *p*-terphenyl (— · — ·).

migrate a long distance in the time, which is very short compared to the fluorescence emission time:

$$S_1^* + S_2 \;\rightleftharpoons\; (S_1 S_2)^* \;\rightleftharpoons\; S_1 + S_2^*$$
$$S_2^* + S_3 \;\rightleftharpoons\; (S_2 S_3)^* \;\rightleftharpoons\; S_2 + S_3^*$$

etc. Each of these reactions is an equilibrium. Thus it is possible to have excimer formation and breaking apart without energy transfer.

A second theory of energy transfer between solvent molecules involves energy migration from one solvent molecule to its adjacent neighbors. This theory is discussed at length by Voltz and co-workers (19):

$$S_1^* + S_2 \;\rightleftharpoons\; S_1 + S_2^*$$
$$S_2^* + S_3 \;\rightleftharpoons\; S_2 + S_3^*$$

etc. Again these energy transfer reactions are equilibriums so that not every contact will lead to energy transfer.

Both of these mechanisms explain the observed evidence of energy transfer between solvent molecules. Solvent–solvent energy transfer takes place in subnanosecond time and can occur over a distance of many molecular diameters. The efficiency of solvent–solvent energy transfer can be reduced by introducing into the system solvents which are called "diluters" (20). The diluter molecules do not participate in the energy transfer process— either as a quencher or as an energy transfer medium; rather, they act to separate the solvent molecules from direct contact with each other, thus proving that the solvent–solvent energy transfer process requires contact. This implies that solvent–solvent energy transfer is in part diffusion controlled (21).

Solvent–solvent energy transfer processes are monoenergetic. The excited state responsible for the energy exchange is most probably the first excited singlet state. However, there is growing evidence that upper excited states can participate in the energy transfer process under certain conditions (22, 10). The energy transfer occurs between the same excited states in the two solvent molecules. A simplified diagram of the mechanism is shown in Fig. II-6, where only the ground (S_0), first excited (S_1), and second excited (S_2) singlet energy levels are depicted.

Solvent–Solute Transfer

As the energy transfers from solvent to solvent molecule it will move from one environment to another. If solute molecules are present at the relatively low concentrations of 3–10 g/liter, it is not likely that the excited solvent molecule will have direct contact with a solute molecule. However, at these low solute concentrations the energy is quantitatively transferred from excited solvent molecules to solute molecules (23, 24).

The energy transfer from solvent to solute is nonradiative: *No* photons are emitted by the solvent molecule and subsequently absorbed by the solute molecules. Energy transfer occurs in 10^{-11} sec, whereas photon emission by the solvent occurs with a decay time of 30×10^{-9} sec. The energy transfer is also not diffusion controlled, which occurs in the order of 10^{-6} sec at these concentrations (12, 25, 26).

The energy transfer occurs by a resonance transfer process. A theory of the interactions of the dipoles of the two molecules has been developed by Förster in several articles (27). The strength of the dipole–dipole interaction has been shown to be related to the degree of the overlap of the fluorescence spectrum of the donor molecule (solvent) and the absorption spectrum of the acceptor molecule (solute). Also, the distance over which the transfer will occur is related to the extinction coefficient of the acceptor

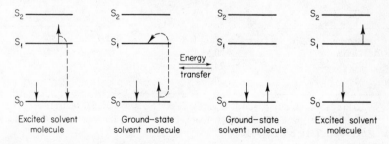

Fig. II-6. Simplified diagram showing energy transfer between two solvent molecules.

molecules over the region of overlap of the two spectra (emission of the donor and absorption of the acceptor).

Förster defined the critical transfer distance R_0 as the distance between an acceptor and a donor molecule such that the probability of transfer is equal to the probability of all other processes of energy release by the donor molecule (radiative emission, quenching, radiationless deactivations, etc.) (28). The rate of energy transfer k when the donor and acceptor molecules are separated by the distance R is given by

$$k = \frac{1}{\tau_0}\left(\frac{R_0}{R}\right)^6,$$

and R_0 can be calculated from the equation

$$R_0{}^6 = \frac{(9000 \ln 10)K^2}{128\pi^5 n^4 N} \int_0^\infty F_D(\bar{v})\, \epsilon_A(\bar{v})\, \frac{d\bar{v}}{\bar{v}^4}$$

where τ_0 is the radiative decay time of the donor molecule, K an orientation factor, n the refractive index of the solvent, N Avogadro's number, $F_D(\bar{v})$ the spectral distribution of the fluorescence emission of the donor molecule, and $\epsilon_A(\bar{v})$ the molar decadic extinction coefficient of the acceptor molecule.

For many molecules the dipole–dipole interaction occurs over intermolecular distances that are large compared to the collision distance. The diameter of most organic molecules is about 6 Å. Two molecules which just contact each other will have a collision distance the sum of the radius of each molecule, or about 6 Å. The Förster radius has been calculated to be 20–60 Å for many of the organic molecules used in liquid scintillation counting.

Table II-4

Förster radius R_0 of PPO and p-terphenyl in some aromatic solvents

Solvent	R_0 of PPO (Å)	R_0 of p-terphenyl (Å)
Benzene	17.5,[a] 16.5[b]	—
Toluene	20.5,[a] 18[b]	27[c]
p-Xylene	24.5[a]	—
Mesitylene	20.5[a]	—

[a] C. L. Braga, M. D. Lumb, and J. B. Birks, *Trans. Faraday Soc.* **62**, 1830 (1966).

[b] R. Voltz, T. Klein, C. Tanielian, H. Lami, F. Heisel, G. Laustriat, and A. Coche, *Coll. Int. Elect. Paris* (31), p. 71 (1963).

[c] S. G. Cohen and A. Weinreb, *Proc. Phys. Soc.* (*London*) **169,** 593 (1956).

Of course the Förster radius R_0 of a solute will vary with the solvent molecule. Table II-4 lists some values of R_0 for PPO and p-terphenyl in different aromatic solvents.

In several recent studies it has been demonstrated that under special conditions it is possible to have the upper excited singlet states of the solvent molecules, $S_2, S_3, \ldots, S_n$, participate in energy transfer processes. At normal solute concentrations (3–10 g/liter) the efficiency is determined by the solvent, benzene $<$ toluene $< p$-xylene $<$ 1,2,3-trimethylbenzene (29). However, at high solute concentrations, at which the energy is transferred from upper excited solvent singlet states, namely S_2 and S_3, the efficiencies of these are the same (10, 22).

Solute–Solute Energy Transfer

Because of the low concentration of the solute it is very improbable that the energy transfer will be diffusion controlled. In fact, because the solute concentrations are low, the radiative probability (i.e., fluorescence) is many times greater than the other types of energy transfer processes (25, 30). There is some energy transfer between like solute molecules which is a long-range interaction process that involves the same energy level in both molecules and is monoenergetic, with essentially no activation energy required.

Energy transfer between unlike solute molecules is also a monoenergetic process but is usually accompanied by a vibrational relaxation in the acceptor molecule which leads to the trapping of the energy in the acceptor molecule. It is possible to have a reverse energy transfer if the acceptor molecule is first given a small amount of excess energy (corresponding to the energy difference in the S_1 excited energy levels of the acceptor and donor molecules) that can be considered as an activation energy. However, this mechanism has not been observed.

Energy Transfer in Liquid Scintillation Solutions

In liquid scintillation counting the excitation energy is in the form of kinetic energy of an ionizing particle produced by a nuclear emission or as a secondary interaction of the materials in the scintillator solution with a nonionizing nuclear emission. Essentially all of the primary excitations result in the formation of excited solvent molecules. The energy then migrates from one solvent molecule to another until the energy is trapped by a solute molecule. If there is a secondary solute, the energy is subsequently transferred from the primary solute to the secondary solute where it is trapped. Finally the energy is released in the form of a photon ($E = h\nu$) which is characteristic of the fluorescent species. In Fig. II-7 is shown a simplified diagram of the many processes that take place in the solution. All of the photons that are emitted from the solution are from the first excited singlet state of the solute. The number of photons per event is proportional to the number of excited molecules initially produced.

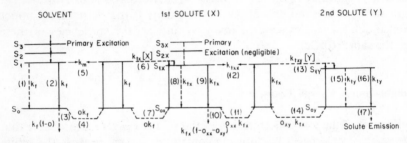

Fig. II-7. Processes of energy transfer in three-component liquid scintillator solution. (1) Internal quenching, (2) emission, (3) escape of light, (4) radiative migration, (5) nonradiative migration, (6) nonradiative transfer to first solute, (7) radiative transfer to first solute, (8) external quenching, (9) emission, (10) escape of light, (11) radiative migration, (12) nonradiative migration, (13) nonradiative transfer to second solute, (14) radiative transfer to second solute, (15) internal quenching, (16) emission, (17) escape of light.

Fluorescence

In Fig. II-8 is shown a modified Jablonski diagram (31) of various processes that can occur when a molecule is excited. The general term *luminescence* is used to describe the emission of light from an excited species; *fluorescence* describes emission from the singlet excited states, usually the S_1 excited state; and *phosphorescence* is the emission from the triplet excited state. Phosphorescence is a forbidden transition (a spin change) and therefore has a longer lifetime. The lifetime is a measure of the time for $1/e$ of the excited molecules to emit photons. The longer the lifetime, the greater is the forbiddenness of the transition. As indicated by the allowed transition (no spin change), fluorescence decay times are very short, of the order of 10^{-5}–10^{-9} sec. Most organic scintillator solutes have decay times of a few nanoseconds (1 nsec = 10^{-9} sec). In liquid media the triplet state emission is not observed. Because of the long life of the triplet state, the triplet energy is usually lost through quenching or triplet–triplet-type processes.

Fluorescence usually occurs between the first excited singlet state (S_1) and the ground state of the given molecule (except in certain cases such as azulene

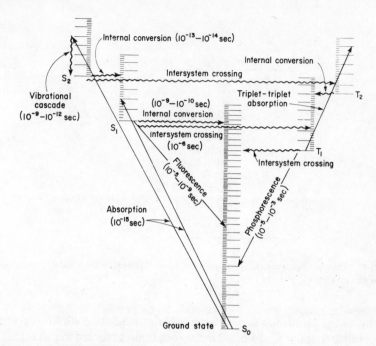

Fig. II-8. Modified Jablonski diagram showing the various processes that can occur upon excitation of an organic molecule: — transitions with adsorption or emmission of radiation; ~ nonradiative transitions.

which has the fluorescence from the S_2 state). The emission is not mono-energetic. The energies of the photons cover a rather wide band. These bands correspond to the energy difference between the zero vibrational level of the S_1 state and the many vibrational levels of the ground state. The types of fluorescence transitions are schematically shown in Fig. II-9.

The fluorescence between $S_{10} \rightarrow S_{00}$ corresponds to the maximum energy of a photon emitted by this excited molecule. Other transitions, $S_{10} \rightarrow S_{0,1,2,...,n}$, have less energy associated with the emitted photon. The remaining energy difference ($E_{S_{10} \rightarrow S_{00}} - E_{S_{10} \rightarrow S_{0n}}$) is dissipated as excess kinetic energy of the ground-state molecule's vibrational energy. Thus the total energy is conserved.

The fluorescence spectrum will reflect the probability of the transition between S_{10} and the vibrational levels of the ground state of the molecule. Most organic scintillator solutes have fluorescence spectra that have several peaks of different intensity. In most cases the $S_{10} \rightarrow S_{00}$ band is either very weak or missing and the probability of the $S_{10} \rightarrow S_{00}$ transition is very small. The transition between S_{10} and the low vibrational levels of the ground state is the most probable. (Because absorption can occur between

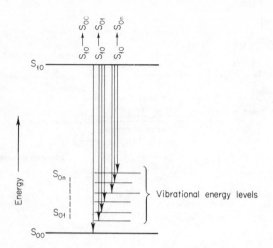

Fig. II-9. Fluorescence transitions between first excited singlet state (S_{10}) and vibrational levels of the ground state (S_{0n}).

the $S_{00} \rightarrow S_{10}$ levels and not between other ground-state vibrational levels, it is possible that the $S_{10} \rightarrow S_{00}$ fluorescence is not observed due to reabsorption of the emitted photon.)

Scintillation Efficiency

An important property of a liquid scintillator solution is its scintillation efficiency S_x. The value of S_x is defined as the ratio of the energy produced as photons by the scintillator solution upon excitation by a particle to the energy of that particle E_{ex}. The total energy produced is the sum of the energy of each photon produced. The energy of a photon is given by Planck's equation

$$E = h\nu$$

where ν is the frequency of the photon and h is Planck's constant. The values of h in some useful units are

$$h = 6.625 \times 10^{-27} \text{ erg-sec} = 4.12 \times 10^{15} \text{ eV-sec.}$$

The relationship between frequency (ν) and wavelength (λ) is

$$\nu = c/\lambda,$$

where c is the velocity of light (2.99×10^{10} cm/sec). Using these equations it is possible to calculate the energy of a photon of wavelength λ. A photon of wavelength 380 nm (1 nm = 10^{-9} m) would have the energy in electron volts (eV) equal to

$$E = (4.12 \times 10^{-5} \text{ eV-sec}) \left(\frac{2.99 \times 10^{10} \text{ cm/sec}}{3.80 \times 10^{-5} \text{ cm}} \right) = 3.25 \text{ eV}.$$

The photons produced as a result of the stoppage of the ionizing particle will have a variation of energies depending on the particular solute and the energy level of the excited fluorescent state. The solute PPO in toluene will emit photons with wavelengths between 330 and 500 nm; the photon energies are between 2.46 and 3.73 eV. The total energy in the form of photons is the energy of each photon times the number of photons of that energy summed over the wavelength spread of the photons. The emission spectrum for a solute is a plot of the wave number ($\bar{v} = 1/\lambda$) versus the relative number of photons of the given wave number. The emission spectrum for PPO in toluene is shown in Fig. II-10. If a very large number of photons were produced, the probability of a photon of a given energy being present would be given by the relative intensity of that wave number.

The total photon probability is proportional to the area under the spectrum. The probability for photons of a small band of wave numbers is the ratio of the area under the spectrum between the wave number limits to the total area. The probability of the photons from excited PPO in toluene having wavelengths between 380 and 390 nm ($\bar{v} = 26{,}316{-}25{,}641$ cm^{-1}) is the ratio of the shaded to the total area (Fig. II-10). The probability is

$$\text{prob(380–390 nm)} = 6/41.5 = 0.145,$$

or about 14.5% of an infinite number of photons will have wavelengths between 380 and 390 nm.

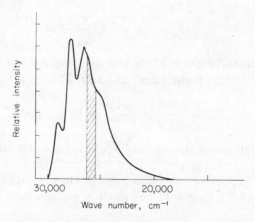

Fig. II-10. The emission spectrum for PPO in tolvene.

The total energy in the form of photons is given by the equation

$$TE_{ph} = h \int_{v_2}^{v_1} n(v) \, dv \tag{1}$$

where the wavelength region of emission is between $v_1 = c/\lambda_1$ and $v_2 = c/\lambda_2$, and $n(v)$ is the number of photons of frequency v. The scintillation efficiency is given by the equation

$$S_x = TE_{ph}/E_{ex}. \tag{2}$$

The actual calculation can be simplified somewhat by a few approximations. First, an average number of photons per unit excitation energy $\bar{n}_{ph}$ is selected. Second, from the emission spectrum, an average energy per photon $\bar{E}_{ph}$ is calculated. Further, the average number of photons for a given excitation energy E_{ex} is given by

$$\bar{N}_{ph} = \bar{n}_{ph} \times \frac{E_{ex}}{\text{unit of energy}}. \tag{3}$$

The scintillation efficiency can be approximated by the substitutions of these average values into Eq. (2) to give

$$S_x = \frac{\bar{N}_{ph} \times \bar{E}_{ph}}{E_{ex}}. \tag{4}$$

The scintillation efficiencies of several organic compounds for excitation by high-energy electrons have been determined experimentally, some of which are summarized in Table II-5. If it is assumed that the scintillation efficiency is 0.04 for a scintillator solution excited by high-energy electrons, the average number of photons produced per unit energy can be calculated:

$$S_x = 0.04 = \frac{\bar{N}_{ph} \times \bar{E}_{ph}}{E_{ex}}. \tag{5}$$

For the scintillator solute PPO the average wavelength of photon emission is 380 nm (32), which is equivalent to 3.2 eV. Therefore

$$\bar{N}_{ph} = \frac{(0.04)(E_{ex} \text{ in eV})}{3.2 \text{ eV}} = 12.5 E_{ex} \text{ (keV)}.$$

Using this relationship, the values of $\bar{N}_{ph}$ have been calculated for a number of different electron energies and are listed in Table II-6. It is shown that even a 1-keV electron will, on the average, produce enough excited molecules to emit 13 photons. Even a 0.1-keV electron will give an $\bar{N}_{ph}$ greater than 1.

Table II-5

Experimentally measured scintillation efficiencies

Anthracene crystals	0.040[a]
Toluene (PPO + POPOP)	0.052[b]
Benzene (*p*-terphenyl)	0.042[c]

[a] J. B. Birks, "The Theory and Practice of Scintillation Counting," pp. 241–244. Pergamon, Oxford, 1964.
[b] J. W. Hasings and G. Weber, *J. Opt. Soc. Amer.* **53**, 1410 (1963).
[c] P. Skarstad, R. Ma, and S. Lipsky, *Mol. Cryst.* **4**, 3 (1968).

When these photons escape from the scintillator solution and are directed onto the face of a multiplier phototube, a certain fraction of them will interact with the photocathode material, producing a certain average number of photoelectrons. The fraction of photons that interact is the *quantum efficiency*. The photocathode quantum efficiency varies with the wavelength of the photons. Considering two types of multiplier phototubes,

Table II-6

Average number of photons and photoelectrons and the expected counting efficiency of a coincidence scintillation counter as a function of the energy of an electron stopped in a liquid scintillator solution

Electron energy (keV)	$\bar{N}_{ph}$	Number of photoelectrons at photocathode efficiency of		Counting efficiency in coincidence counter with photocathode efficiency of	
		15%	28%	15%	28%
1000	12,500	1875	3500	100	100
500	6250	938	1750	100	100
158	1980	297	555	100	100
50	625	94	175	100	100
5	63	9	18	99.6	100
1	13	2 (86%)[a]	3.5 (97%)[a]	37[b]	77[c]
0.1	1.3	0.2 (15%)[a]	0.4 (27%)[a]	0	0

[a] This is the percentage of the events which produced the noted number of photoelectrons that will produce a measurable pulse, i.e., a count.

[b] This efficiency is calculated from the probability of coincidence and the square of the Poisson probability of the MPT:

$$0.50 \times (0.86)^2 = 0.37.$$

[c] Using same procedure as in the example above:

$$0.82 \times (0.97)^2 = 0.77.$$

one with an average quantum efficiency of 15% and a second with an average quantum efficiency of 28%, it is possible to calculate the average number of photoelectrons produced by $\overline{N}_{ph}$. Table II-6 lists the calculated photoelectron production for these two cases, first assuming that all of the photons fall on a single multiplier phototube, and second, assuming that they are divided between the photocathodes of two multiplier phototubes in coincidence.

When the number of photons is small, it is necessary to introduce the statistical probabilities governing small numbers. In a multiplier phototube, the statistics of small numbers follows Poissonian probability. With modern-day multiplier phototubes, a single photoelectron will, on the average, produce enough amplification to give a measurable current at the anode, which is recorded as a count. However, each photoelectron multiplication process is subject to the statistical processes at each dynode of the multiplication system.

There is a finite probability that any number of photoelectrons will not produce a measurable pulse at the anode. This probability becomes increasingly small as the number of photoelectrons in the pulse increases. Also, since a single photon can produce a photoelectron, there is a finite probability that events which on the average produce insufficient photons to produce a measurable pulse will produce a measurable pulse. As a result of these statistical factors, a part of all events that produce small numbers of photons will fail to be detected, with the final result reflected in a lowering of counting efficiency. Table II-7 lists the Poisson statistics

Table II-7

Probability that a scintillation event which produces a given number of photoelectrons will be measured

Number of photoelectrons	Probability of measurable[a] pulse (%)	Probability of nonmeasurable pulse (%)
1	63.3	36.7
2	86.5	13.5
3	95.0	5.0
4	98.2	1.8
5	99.3	0.7
⋮	⋮	⋮
10	99.995	0.005

[a] Assuming that a single photoelectron is measurable if it produces at least the average gain in the MPT.

which give the probability of measuring events that produce a given number of photoelectrons.

Counting Efficiency

The scintillation efficiency of scintillator solutions should not be confused with counting efficiency. The *counting efficiency* is a measure of the fraction of particles that produce a measurable pulse (i.e., a count). Alpha-particle excitation of a liquid scintillator gives a scintillation efficiency of only 0.4% but a counting efficiency of 100% (33). An excitation event can convert a very small fraction of the particle energy to photons, and yet if that number of photons is sufficiently large, there is essentially a 100% probability that it will produce a measurable pulse.

However, this is not to say that there is not a direct relationship between scintillation efficiency and counting efficiency, for there is. The scintillation efficiency will determine the number of photons produced, on the average, per unit energy of the given type of particle that excites the solution. The scintillation efficiency is affected by the composition of the solution (solvent, solutes, quenchers), whereas the counting efficiency reflects not only the scintillation efficiency but the optical properties of the system, the properties of the multiplier phototube (quantum efficiency, multiplication, spectral response, etc.), and the electronic recording (threshold, pulse height analyzer, etc.).

The requirement of coincidence is one factor which limits the counting efficiency for low-energy particles. Coincidence measurement requires that a measurable pulse be obtained in each of two multiplier phototubes within the resolving time of the system (20–30 nsec). However, this requires that the photons be divided between the two multiplier phototubes. If the quantum efficiency of the multiplier phototubes is 28%, 3.5 photons would be required on the average to produce a single photoelectron. However, in a coincidence system twice this number of photons is necessary to produce a measurable pulse, since a photoelectron has to be produced in each multiplier phototube.

Consider an excitation that produces seven photons: What is the probability that a measurable pulse will be obtained? This probability will be the result of several different probabilities:

(a) division of the photons between the two multiplier phototubes,
(b) probability of producing a photoelectron,
(c) probability of gain necessary to give a pulse of amplitude greater than threshold.

There are other probabilities that are also important but which can be considered to be nearly unity or at least a constant factor, for instance, optical system for photons, transit time of multiplier phototube, etc.

The division of the photons can be calculated from the equation (34):

$$\text{probability} = 1 - 2^{(1-M)}$$

where M is the expected number of photoelectrons; for a coincidence system M has to be greater than 2. Table II-8 lists the probability for several values of M.

Table II-8

Probability of a coincidence event as function of the expected average number of photoelectrons due to a scintillation event

Expected number of photoelectrons, M	Probability
<2	0
2	0.50
3	0.75
4	0.875
5	0.9375
⋮	⋮
10	0.99805

The probability that a photoelectron is produced at the photocathode is given to be Poisson probability. The probability that at least one photoelectron is produced for an expected single photoelectron (3.5 photons striking the photocathode) is 0.633, or about 0.367 of the time a photoelectron will not be produced.

Finally, if at least one photoelectron is produced, the probability that it will produce an anode pulse greater than the threshold (i.e., the average gain of the multiplier phototube) is 0.982.

The probability that a coincidence pulse will be obtained is given by

$$P = (0.50)(0.633)^2(0.982)^2 = 0.193$$

where the photocathode probability (0.633) and the gain probability (0.982) are squared because these occur in each multiplier phototube independently. Thus there is only a 19.3% probability that an event producing seven photons will be measured in a coincidence counter.

In a single multiplier phototube counter, with all the photons directed on the photocathode, the same event of seven photons will be measured with a probability of

$$P = (0.865)(0.9997) = 0.865,$$

or 86.5% of the time this event produces a measurable pulse. However, the background will also be very high.

Using this type of calculation the counting efficiencies of several isotopes in coincidence counters have been calculated for various values of the average energy necessary to produce a measurable pulse (35, 36). The calculated efficiencies are listed in Table II-6.

References

1. J. B. Birks, *Proc. Phys. Soc. London Sect. A* **64,** 874 (1951).
2. J. K. Basson and J. Steyn, *Proc. Phys. Soc. London Sect. A* **67,** 297 (1954).
3. H. H. Seliger, *Int. J. Appl. Radiat. Isotop.* **8,** 29 (1960).
4. D. L. Horrocks, *Rev. Sci. Instrum.* **35,** 334 (1964).
5. K. F. Flynn, L. E. Glendenin, E. P. Steinberg, and P. M. Wright, *Nucl. Instrum. Methods* **27,** 13 (1964).
6. R. D. Brooks, *Progr. Nucl. Phys.* **5,** 284 (1956).
7. D. L. Horrocks, *Rev. Sci. Instrum.* **34,** 1035 (1963).
8. A. Skerbele and E. N. Lassettre, *J. Chem. Phys.* **42,** 395 (1965).
9. H. B. Klevens and J. R. Platt, *J. Chem. Phys.* **17,** 470 (1949).
10. D. L. Horrocks, *J. Chem. Phys.* **52,** 1566 (1970).
11. J. B. Birks, C. L. Braga, and M. D. Lumb, *Brit. J. Appl. Phys.* **15,** 399 (1964).
12. D. L. Horrocks, *Photochem. and Photobiol.* **15,** 239 (1972).
13. E. Langenscheidt, *Nucl. Instrum. Methods* **91,** 237 (1971).
14. G. T. Reynolds, *Nucleonics* **10** (7), 46 (1952).
15. M. Furst and H. Kallman, *Phys. Rev.* **85,** 816 (1952).
16. F. N. Hayes, B. S. Rogers, P. Sanders, R. L. Schuch, and D. L. Williams, Rep. LA-1639, Los Alamos Sci. Lab., Los Alamos, New Mexico, 1953.
17. J. B. Birks, J. C. Conte, and G. Walker, *IEEE Trans. Nucl. Sci.* **NS-13** (3), 148 (1966).
18. J. B. Birks, *in* "Organic Scintillators and Liquid Scintillation Counting" (D. L. Horrocks and C. T. Peng, eds.), p. 3. Academic Press, New York, 1971.
19. R. Voltz, G. Laustriat, and A. Coche, *C. R. Acad. Sci. Paris* **257,** 1473 (1963).
20. F. N. Hayes, B. S. Rogers, and P. C. Sanders, *Nucleonics* **13** (1), 46 (1955).
21. A. Weinreb, *Proc. Conf. Organic Scintillation Detectors, Univ. of New Mexico, 1960,* (G. H. Daub, F. N. Hayes, and E. Sullivan, eds.), TID-7612, p. 59. U. S. At. Energy Commission, Washington, D. C., 1961.
22. G. Laustriat, R. Voltz, and J. Klein, *in* "The Current Status of Liquid Scintillation Counting" (E. D. Bransome, Jr., ed.) p. 13. Grune & Stratton, New York, 1970.
23. S. Lipsky and M. Burton, *J. Chem. Phys.* **31,** 1221 (1959).
24. I. B. Berlman, *J. Chem. Phys.* **33,** 1124 (1960).
25. J. B. Birks and K. N. Kuchela, *Proc. Phys. Soc. (London)* **77,** 1083 (1961).
26. D. L. Horrocks, Unpublished results, 1971.
27. See review article by Th. Förster, *Radiat. Res.* **2,** 326 (1960).
28. Th. Förster, *Discuss. Faraday Soc.* **27,** 7 (1959).
29. C. W. Lawson, F. Hirayama, and S. Lipsky, *J. Chem. Phys.* **51,** 1590 (1969).
30. D. L. Horrocks, *in* "Organic Scintillators" (D. L. Horrocks, ed.), p. 45. Gordon & Breach, New York, 1968.
31. A. Jablonski, *Z. Phys.* **94,** 38 (1935).
32. I. B. Berlman, "Handbook of Fluorescence Spectra of Aromatic Molecules," p. 148. Academic Press, New York, 1965.

33. D. L. Horrocks, *Prog. Nucl. Energy Ser.* 9 **7,** 21 (1966).
34. R. K. Swank, *in* "Liquid Scintillation Counting" (C. G. Bell and F. N. Hayes, eds.), p. 23. Pergamon, Oxford, 1958.
35. D. L. Horrocks and M. H. Studier, *Anal. Chem.* **33,** 615 (1961).
36. J. A. B. Gibson and H. J. Gale, *J. Sci. Instrum.* **1,** 99 (1968).

CHAPTER III

SCINTILLATOR SOLUTIONS

The scintillator solution (excluding the sample) is composed of a solvent (or solvents) and a solute (or solutes). The solvent acts as a medium for absorbing the energy of the nuclear radiation and for dissolving the sample. The solute acts as an efficient source of photons after accepting energy from the excited solvent molecules.

Solvents

Most often the solvent is thought of as the vehicle for dissolving the sample and solutes, and it is often an overlooked fact that the initial excitations occur in the solvent molecules. Thus it is necessary that a solvent be selected which is efficient for the production of excited states upon interaction with the nuclear radiations. The type of molecules that produce excited states easily are those which have available low-lying energy levels and nonbonding electrons (π) which require little energy for promotion into the higher energy levels. These energy levels have to be long-lived relative to the time of migration of the excitation between molecules.

The aromatic-type molecules usually meet the energy and π-electron requirements. The simple aromatic molecules are also liquids at room temperature, giving rise to their use as liquid scintillator solvents.

Solvents are usually divided into three classes, depending on their scintillation efficiency. These classes are

(a) effective solvents—most aromatic compounds;

(b) moderate solvents—many saturated hydrocarbons with scintillation efficiencies between 15 and 50% of the aromatics; and

(c) poor solvents—compounds such as alcohols, esters, ketones, etc., with scintillation efficiencies less than 1% of the aromatics.

Hayes *et al.* (1) investigated the relative scintillation response of many solvents relative to toluene. Some of their results are shown in Fig. III-1. In each case successively increasing amounts of the solvent were added to toluene with a scintillator solute (PPO at 3 g/liter). The response was measured by the relative pulse height produced by the 630-keV conversion electrons from a $^{137}Cs-^{137m}Ba$ radioactive source.

Some of the solvents showed very rapid decrease in the relative pulse height response, indicating a strong interaction with excited molecules produced in the toluene system. Such solvents were bromobenzene, bromocyclohexane, 3-picoline, piperidine, and pyridine. Other solvents showed only a slight decrease in the relative pulse height response, indicating that those solvents were nearly as efficient as toluene. Such solvents were fluorobenzene and *o-, m-,* and *p-*fluorotoluene. The rest of the compounds were intermediate in their effect. This latter group of solvents probably acted as diluters, i.e., they diluted the excited states, thus reducing the efficiency of the excitation energy migration between solvent molecules.

Solvents are also classified as primary and secondary, depending on the relative amounts and their functions in the scintillation processes. The primary solvent is the initial energy absorber and produces the initial excited molecules; the secondary solvent will act as an intermediary in the energy transfer process, which increases the efficiency of the energy migration from initial primary solvent excited molecules to the emitting solute molecules. Naphthalene acts as such an intermediary in the energy transfer processes in solutions with dioxane as the primary solvent.

The primary solvents used today are essentially the same ones that were first used in the early studies of 1949–1955. Since 1955, there have been no new scintillator solvents reported. However, the availability of good scintillation-grade solvents at reasonable prices has improved the scintilla-

ORGANIC LIQUID SCINTILLATORS

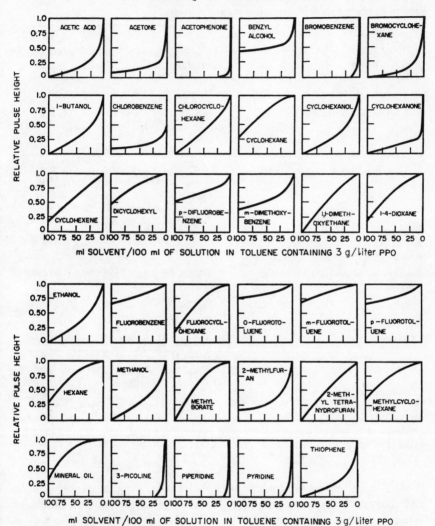

Fig. III-1. Relative scintillation efficiencies of mixtures of organic solvents and toluene (1).

tion efficiency of solutions and, more important, has reduced the variations in impurity content from one batch to another.

The best solvents are the aromatic solvents. Among these there is also a variation in scintillation efficiency. Table III-1 lists some of the commonly used solvents and the relative scintillation efficiency of the solvent with 3 g/liter of PPO.

Table III-1

Relative scintillation yields of some commonly used solvents

Solvent	Structure	Relative scintillation yield[a]
1,2,4-Trimethylbenzene	H_3C—benzene ring with CH_3 (top) and CH_3	112
p-Xylene	H_3C—benzene ring—CH_3	110
Toluene	H_3C—benzene ring	100
Benzene	benzene ring	85
Dioxane	ring: CH_2—CH_2 / O ... O / CH_2—CH_2	65
Cyclohexane	ring with S	20

[a] Measured by relative (toluene = 100) pulse height of the Compton edge for the 662-keV γ rays of ^{137m}Ba.

Most investigators use toluene as the solvent for organic soluble samples and dioxane for aqueous samples. However, with the modern emulsifier systems, aqueous solutions can be counted with very high efficiency in aromatic solvent solutions. The availability of toluene in high purity at moderate prices has led to its use over p-xylene and 1,3,4-trimethylbenzene, which have higher scintillation efficiencies.

Dioxane is still used by many investigators for counting aqueous solutions. There are many problems with maintaining dioxane of high purity. The dioxane undergoes decomposition to form among its products such compounds as peroxides, which are efficient quenchers of the excited molecules. The shelf life of dioxane is somewhat dependent on the impurities present, which seem to accelerate the decomposition rate. Distillation of the dioxane and storage in amber bottles under inert atmosphere (nitrogen or argon) can increase the shelf life.

Because of the problems with dioxane purity, many investigators have looked for other means for efficiently counting aqueous solutions and water. The newly developed emulsifier systems have been very successful in filling this need. These systems are discussed in detail in Chapter VI.

The scintillation efficiency will be determined by the solvent used under the most ideal conditions. This is true because:

(a) the concentration of solutes can be chosen such that the energy transfer efficiency is 100%,

(b) solutes can be chosen which have close to 100% quantum efficiency (fraction of excited molecules that emit a photon),

(c) systems can be chosen which have no quenching of excited molecules by foreign materials, and

(d) solutes can be selected that emit photons of energy (wavelength) which have the maximum detection probability (escape from the solution and interaction with the photocathode of the multiplier phototube).

The scintillation efficiency of solutions with aromatic solvents is near 4% for electron excitation. The efficiency is measured as the ratio of the energy (as photons) emitted from the solutes to the energy of the exciting electron. Lipsky and co-workers (3, 4) in recent studies investigated the possible difference between the efficiencies of different solvents. It was shown (3) that the efficiency of internal conversion from upper excited states of the solvent molecules $(S_2, S_3, \ldots, S_n)$ to the first excited singlet state of the solvent (S_1) is not the same for the various aromatic solvents. Their measurements are summarized in Table III-2. The rating of the solvents in Tables III-1 and III-2 is not the same in absolute values, but the general trend is the same. This is due to the fact that only about 25–40% of the final S_1 excited state is produced by internal conversion from upper excited

Table III-2

Internal conversion efficiency of benzene, toluene, and *p*-xylene as reported by Lawson *et al.*[a]

Solvent	Efficiency of $S_3 \rightarrow S_1$ internal conversion
Benzene	0.45
Toluene	0.76
p-Xylene	1.03

[a] C. W. Lawson, F. Hirayama, and S. Lipsky, *J. Chem. Phys.* **51,** 1590 (1969).

Table III-3

Calculation of total yield of S_1 state of different aromatic solvents and comparison with the measured relative scintillation yields at solute concentrations of 5 g/liter

Solvent	Yield of S_1 from ion recombination	Yield of S_1 from internal conversion of S_3	Total yield of S_1		Measured scintillation yield[b] at solute concentration of 5 g/liter		
			G value	Percent[a]	1,1'-BN	i-p-PBD	PPO
Benzene	1.2	$0.44 \times 0.9 = 0.4$	1.6	76	80	78	78
Toluene	1.2	$0.76 \times 0.9 = 0.7$	1.9	90	92	91	89
p-Xylene	1.2	$1.00 \times 0.9 = 0.9$	2.1	100	95	97	98
1,2,4-Trimethylbenzene	1.2	$1.00 \times 0.9 = 0.9$	2.1	100	100	100	100

[a] Normalized to 100 for maximum possible yield, namely 100% internal conversion of $S_3 \to S_1$.
[b] Normalized to 100 for maximum measured scintillation yield, namely that for 1,2,4-trimethylbenzene.

states (5). The majority of S_1 excited states are produced directly by ion recombination (6–8).

Table III-3 shows the agreement between the calculated and the measured scintillation efficiencies for the aromatic solvents benzene, toluene, p-xylene, and 1,2,4-trimethylbenzene (5). The agreement is very good, which seems to substantiate the theory of nonquantitative internal conversion of some solvents. Similar results obtained by Birks (8) are shown in Table III-4.

The internal conversion efficiency of the solvent is important in the determination of the relative scintillation efficiency, because only a very small fraction of the excited state produced by the ionizing radiation is in the S_1 state. The majority of excited states are higher-energy excited states, namely, S_2 and S_3. Table III-5 lists the relative yields of the first three excited states of benzene, toluene, p-xylene, and 1,2,4-trimethylbenzene (9, 10).

Some of the scintillation properties of solvents are listed in Table III-6. The reasons that solvents alone are not good scintillation systems are obvious from these data. It is necessary to have a solute present because:

(a) solvents usually have poor quantum yields (Φ),

(b) the wavelength distribution of solvent fluorescence is in a range (200–300 nm) where the multiplier phototubes are not very sensitive,

(c) a large amount of self-absorption will occur even in moderate volumes, due to the overlap of the absorption and emission spectra, and

(d) the long decay times (~ 30 nsec) are so long that quenching (nonradiative deexcitation) can be competitive.

The purity of the solvents is most often very important. Small amounts of some compounds can lead to almost complete loss of scintillation efficiency of the solution. Also, those solvents that have isomers which are less efficient can give apparent differences in scintillation efficiency due to the presence of these isomers. In the case of xylenes, p-xylene is a much more efficient scintillator than either o-xylene or m-xylene. Therefore, the presence of these isomers can affect the scintillation efficiency of the solution (Table III-7).

In several recent studies (5, 11–15) there has been some evidence to indicate that certain quenching processes and energy transfers can involve the upper excited states of the solvent molecules, rather than exclusively the first excited singlet state S_1. Some workers showed that certain quenchers will preferentially accept energy from the upper excited states of solvent molecules such as toluene (12–14). The molecule $CHCl_3$ will lead to quenching of excitation when ionizing radiation causes the excitations but will not quench when uv light is used to produce only the S_1 excited states of toluene. The ionizing radiations will produce S_3, S_2 states, which can transfer energy to $CHCl_3$.

Table III-4

Properties of liquid scintillator solvents excited by fast electrons[a]

Solvent	Excitation				Ion recombination		Total		Expt.	
	S_E (kk)	$G_0(S_E)$	$C(S_E)$	$G_{0E}(S_{1x})$	I_x (eV)	$G_{0I}(S_{1x})$	$G_0(S_{1x})$	$g_0(S_{1x})$	$g_0(hvy)$ PPO[b]	PPO[c]
Benzene	54.3	0.9	0.45	0.4	9.25	1.2	1.6	82	85	87
Toluene	53.0	0.92	0.76	0.7	8.81	1.26	1.96	100	100	100
m-Xylene	51.3	0.95	1.00	0.95	8.59	1.29	2.24	114	109	—
p-Xylene	51.5	0.95	1.00	0.95	8.44	1.32	2.27	116	112	110
1,2,4-Trimethylbenzene	51.0	0.96	1.00	0.96	8.27	1.34	2.30	118	—	112
Isodurene	50.9	0.96	1.00	0.96	8.03	1.39	2.35	120	—	—
1-Methylnaphthalene	44.6	1.10	1.00	1.10	7.96	1.39	2.49	127	—	—

[a] See J. B. Birks, *in* "Organic Scintillators and Liquid Scintillation Counting" (D. L. Horrocks and C. T. Peng, eds.), p. 3. Academic Press, New York, 1971.

[b] 3 g/liter PPO (F. N. Hayes *et al.*, *Nucleonics* **13**, (1), 46 (1955).)

[c] 5 g/liter PPO (D. L. Horrocks, *J. Chem. Phys.* **52**, 1566 (1970).)

Table III-5

**Relative yields of excited states of aromatic solvents benzene, toluene,
p-xylene, and 1,2,4-trimethylbenzene[a]**

Excited state	Relative yield	Normalized yield (%)
S_1	0.002	0.3
S_2	0.10	12.6
S_3	0.69	87.1

[a] As predicted by optical approximation model.

In a different study on the role of upper excited states in the energy transfer process, it was shown that the scintillation efficiency of solvents could be altered by further increases in the solute concentration beyond the normal "100% energy transfer" concentration (5). The scintillation efficiencies of benzene, toluene, *p*-xylene, and 1,3,4-trimethylbenzene are all equal at very high concentrations of such solutes as 1,1'-binaphthalene, butyl-PBD, and PPO. The conclusion was that at these high concentrations the energy transfer from solvent to solute occurred between the upper excited states of the solvent and solute, i.e., $S_3 \rightarrow F_3$, $S_2 \rightarrow F_2$, where S denotes solvent molecule and F denotes solute molecule.

However, in liquid scintillation counting solutions at normal solute concentrations (3–10 g/liter) the excited solvent molecules will undergo

Table III-6

Fluorescence properties of solvents used in liquid scintillation counting

	Absorption λ (nm)	Emission λ (nm)	Quantum yield	Decay time (nsec)	Internal conversion $S_3 \rightarrow S_1$
Benzene	255	283	0.06	29	0.45
Toluene	263	285	0.14	34	0.76
p-Xylene	266	291	0.34	30	1.00
1,2,4-TMB	270	293	0.33	27.2	1.00
p-Dioxane (neat)	185	247	0.03	2.1	—
p-Dioxane + H$_2$O (5%)	185	325	0.01$_5$	—	—
1-Methylnaphthalene	280	338	0.21	67	—
2-Ethylnaphthalene	280	330	0.26	50	—
m-Xylene	265	289	0.14	30.8	—
o-Xylene	265	289	0.16	32.2	—
Anisole methoxybenzene	270	296	0.26	27.4	—

Table III-7

Xylene scintillation efficiencies

Solvent	Structure	Scintillation efficiency
p-Xylene	H_3C—⟨benzene ring⟩—CH_3	0.33
m-Xylene	H_3C—⟨benzene ring with CH_3⟩	0.14
o-Xylene	H_3C—⟨benzene ring with CH_3⟩	0.16

internal conversion and other deexcitation processes until all the excited singlet molecules are in the first excited singlet energy level. The energy migration and transfer processes occur between the solvent and solute first excited singlet states—i.e., $S_1 \rightarrow F_1$.

One other process that occurs in the solvent excited molecule is called *intersystem crossing*. This process is used to describe the change between the two excitation manifolds that an excited molecule can have, the singlet and the triplet excited states. The triplet state occurs when the electron excited in the energy absorption process has a change in its spin. Normally, the molecule has all paired electrons, i.e., two electrons of opposite spin. When energy is absorbed it can be considered (for a simplistic view) to be localized upon a single electron. If the spin of the electron is unchanged in the excited molecule, the excited molecule is in a singlet state (assuming the ground state of the molecule is a singlet state (see Fig. III-2). However, if the spin of the excited electron is reversed, i.e., the total spin of the molecule is no longer zero, the molecule is in a triplet state (Fig. III-3).

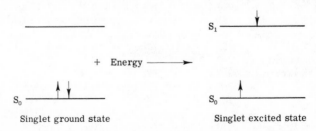

Fig. III-2.   Reaction leading to a singlet-state excited molecule.

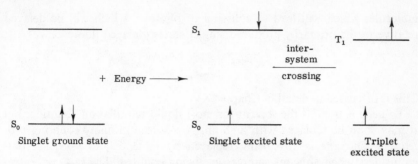

Fig. III-3. Reaction leading to a triplet-state excited molecule.

Because processes that require a spin change are highly forbidden, it is not very probable that an excited triplet state can be directly produced from a singlet ground state. First an excited singlet state is formed which will undergo an intersystem crossing process to give an excited triplet state. It is also possible for upper excited triplet states to undergo intersystem crossing to give excited singlet states.

The triplet states, because of the nature of the forbidden transition from excited triplet states to a singlet state, are usually longer lived than the singlet state. This is especially true when comparing the S_1 and T_1 states (S_1 is the lowest excited singlet state and T_1 is the lowest excited triplet state). Both of these states return to the ground state, at least part of the time, by the emission of photons. The $S_1 \rightarrow S_0$ transition is known as fluorescence and is an allowed transition. For most scintillator molecules the decay times of fluorescence are in the nanosecond time scale. The $T_1 \rightarrow S_0$ transition is known as phosphorescence and is a forbidden transition. The decay times for phosphorescence are much longer than for fluorescence, usually between 10^{-5} and several seconds. Some of these processes are summarized in Fig. II-6.

In most liquid scintillation solutions the triplet state formation is not important. If oxygen or other quenchers are present in the solution, all triplet states will be quenched. The only effect of the yield of triplet states (at least that part formed by intersystem crossing) is to remove excited molecules from the singlet manifold, which will lead to a decrease in scintillation yield. Heavy atoms have been shown to increase the intersystem crossing efficiency (16). Thus if the scintillator sample has molecules with heavy atoms (i.e., iodine, mercury, etc.), the scintillation yield can be decreased due to an increase in the formation of triplet excited molecules.

The yield of triplet states is important in one type of liquid scintillation counting: pulse shape discrimination. If oxygen and other quenchers are excluded from the scintillator system, it is possible to observe the result of two excited triplet molecules reacting and producing a singlet excited molecule, which will lead to emission of photons, which will be delayed relative to the primarily produced singlet states (delayed fluorescence):

$$T_1 + T_1 \rightarrow S_1 + S_0$$
$$S_1 \rightarrow S_0 + h\nu.$$

This is discussed in detail in Chapter XV.

Toluene is used as the solvent for most liquid scintillation counting, but it may soon be replaced with high-purity p-xylene in many counting procedures. Dioxane is also being replaced by toluene (and p-xylene) plus emulsifiers when aqueous samples are being counted. The salt concentration of the aqueous samples is very important in determining which solvents are to be used.

Solutes

Some of the commonly used primary solutes are listed in Table III-8 along with some of their scintillation and chemical properties.

Since most solvents are not efficient scintillators by themselves, it is common to add to the solutions certain organic molecules (solutes) which are highly efficient fluors. These molecules act as acceptors of the energy that is present in the form of excited solvent molecules. In accepting the energy the solute molecules are excited to one of their many excited states (Fig. III-4). Since the excitation energy of the solute molecule is somewhat less (a fraction of an electron volt in most molecules) than the energy of the excited solvent molecule, the energy exchange is not reversible. Thus the energy is trapped by the solute molecules. The excited solute molecules will then return to the ground state with the release of energy. When this energy appears in the form of a photon the process is referred to as fluorescence. The energy can also be utilized in nonphoton-producing events—

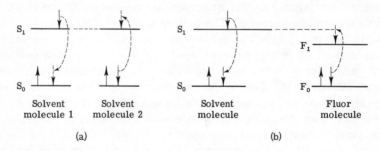

Fig. III-4. Schematic of (a) solvent–solvent energy transfer (monoenergetic), and (b) solvent–solute energy transfer (exothermic).

called nonradiative processes. The fraction of excited molecules which emits photons is called the fluorescence yield Φ:

$$\Phi = (\text{number of photons})/(\text{number of excited molecules}).$$

This equation is valid because each excited molecule will emit only one photon with energy corresponding to the transition between the excited state and one of the many vibrational levels of the ground state.

As seen in Table III-8 most of the commonly used scintillator solutes have high fluorescence yields. The difference between 1.0 and the fluorescence yield is a measure of the nonradiative processes of triplet formation (intersystem crossing), nonradiative deactivation (internal conversion), etc.

Concentration Dependency

The efficiency of the transfer of energy from the solvent to the solute (as discussed in Chapter II) is dependent on the solute concentration. Figure III-5 shows the relative scintillation yield of various solutes as a function of their concentration in toluene as measured by the relative pulse height of the Compton edge produced by the 662-keV gamma (γ) rays from a $^{137}Cs-^{137m}Ba$ source.

With many of the solutes used in liquid scintillation counting the energy transfer process approaches 100% efficiency at moderate solute concentrations, 7–10 g/liter. Berlman (17) showed that essentially all the energy transfer occurred by nonradiative processes. Three different and independent methods gave identical results: measure of the fluorescence lifetime, measure of solvent fluorescence intensity, and measure of solute

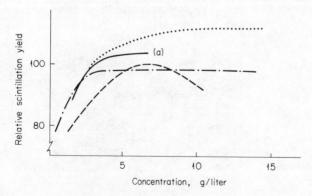

Fig. III-5. Concentration dependency of the scintillation yield. (a) Solubility limit of *p*-terphenyl; — — — PPO; . . . . *t*-butyl-PBD; —— *p*-terphenyl; · — · — · TMQP.

Table III-8
Properties of a few primary solutes

Solute	Peak fluorescence (nm)	Decay time, τ (nsec)	Quantum yield, Φ	Solubility in toluene (g/liter) at room temperature	Optimum solute concentration (g/liter)
p-Terphenyl	342	1.0	0.77	5	5
PPO	375	1.4	0.83	400	3–7
PBD	375	1.0	0.69	12	8–10
Butyl-PBD	385	1.0	0.69	130	12
Naphthalene	334	96.0	0.19	250	[a]

Table III-8 (cont.)

Solute	Peak fluorescence (nm)	Decay time, τ (nsec)	Quantum yield, Φ	Solubility in toluene (g/liter) at room temperature	Optimum solute concentration (g/liter)
9,10-Diphenyl-anthracene	428	9.4	0.83	>35	[a]
BBOT	446	1.6	0.61	60	7

[a] Not normally used as primary solute. Included for comparison.

Table III-9

The efficiency of energy transfer (%) from excited solvent molecules (*p*-xylene)
to solute molecules (PPO) as a function of solute concentration (17)

Concentration of PPO (g/liter)	From lifetime measurements	From *p*-xylene emission intensity	From PPO emission intensity
0.02	14 ± 3	12 ± 2	14 ± 2
0.04	26 ± 3	25 ± 2	23 ± 2
0.10	40 ± 3	39 ± 2	41 ± 2
0.20	55 ± 3	59 ± 2	56 ± 2
0.25	59 ± 3	62 ± 2	62 ± 2
0.50	75 ± 3	80 ± 2	77 ± 2
1.00	84 ± 3	88 ± 2	85 ± 2
2.00	90 ± 3	—	92 ± 2

fluorescence intensity. Table III-9 lists the results. At concentrations greater than 2 g/liter of PPO in *p*-xylene, the energy transfer was more than 90% efficient.

Ideally the solute concentration is chosen to give a maximum of energy transfer. However, it is necessary to have an excess of solute over the minimum required for maximum energy transfer. This excess will allow for any dilutions which occur when introducing the sample or secondary solvents. At the concentration C_B shown in Fig. III-6 it can be readily seen that dilutions (up to a certain degree) will not alter the relative scintillation yield, and thus the counting efficiency will remain constant. However, if the initial concentration were C_A, small dilutions would decrease the relative scintillation yield. For higher-energy beta emitters, such as ^{14}C

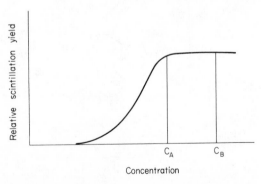

Fig. III-6. Choice of proper solute concentration (C_B).

and ^{32}P, the counting efficiency may not change much. But for low-energy beta emitters, such as ^{3}H, the counting efficiency will be noticeably decreased upon dilution.

However, three major problems occur with selecting the optimum solute concentration: One is the solute solubility in the scintillator solvent; the second is the process of solute self-quenching through solute–solute interactions; and the third arises from self-absorption of the emitted photons due to an overlap of absorption and emission spectra of the solute.

Some organic compounds which would be very efficient scintillator solutes can not be used because their solubility limit occurs before the maximum energy transfer is reached. Figure III-7 shows some compounds that are solubility limited as primary solutes. (As is seen later, some of these compounds have found use as secondary solutes.)

Solute self-quenching occurs when an excited solute molecule interacts with a like ground-state solute molecule by direct contact. The excitation energy is shared by the two molecules. Any of the following processes may occur as a result of the solute–solute interaction:

$$\text{interaction:} \quad F_1^* + F_0 \rightarrow (F_1^* \cdot F_0)$$

(a) energy transfer to
return to initial conditions: $(F_1^* \cdot F_0) \rightarrow F_1^* + F_0$

(b) self-quenching: $(F_1^* \cdot F_0) \rightarrow 2F_0 + en_1$

(c) excimer formation: $(F_1^* \cdot F_0) \rightarrow F_2^* + en_2$

(d) excimer emission: $F_2^* \rightarrow 2F_0 + h\nu_D.$

Process (a) leads to no change in the scintillation yield as there is no decrease in the number of excited molecules, whereas process (b) leads to a decrease in the scintillation yield by decreasing the number of excited molecules. In the selfquenching process the energy is shared by the two molecules. When they separate each molecule will possess some energy,

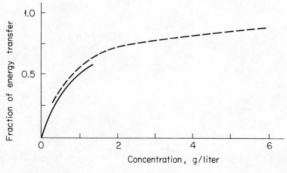

Fig. III-7. Solubility limit effects upon obtaining 100% energy transfer from toluene: — — — *p*-terphenyl; —— M_2-POPOP.

but less than what is necessary to produce a photon. Thus the excess energy is dissipated as heat, or vibrational energy. Process (c) leads to a new form of excitation, an excited dimer (18–21). However, when the excimer fluoresces, the photons are less energetic than the photons from the monomer. Thus the distribution of light is different and the multiplier phototube response may be different. If the multiplier phototube response is less, the scintillation efficiency will be decreased. Also the excimer is quenched to a higher degree than the monomer because of its longer lifetime. Thus in air-saturated scintillator solutions the excimers are efficiently quenched, which leads to a decreased scintillation yield for compounds that form excimers.

The third problem associated with solute concentrations, self-absorption, is critical in the measure of the scintillation yield. Figure III-8 shows the absorption and emission spectra of three scintillator solutes (22). The amount of self-absorption is related not only to the solute concentration but to the molar extinction coefficient ϵ also. The self-absorption is greater with (a) greater overlap, (b) greater concentrations, and (c) higher values of ϵ. Also an important factor in the amount of self-absorption is the path

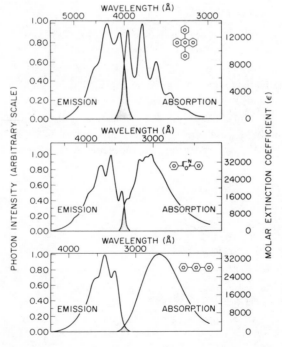

Fig. III-8. Ultraviolet absorption and emission spectra of three commonly used scintillator solutes (22) in cyclohexane.

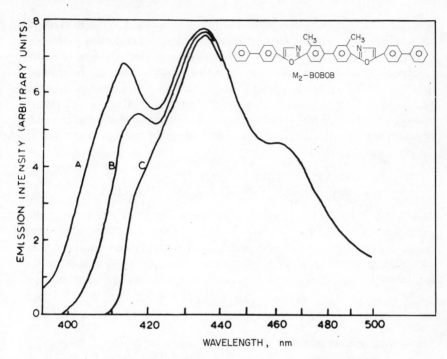

Fig. III-9. Fluorescence spectra of M$_2$-BOBOB upon passage through (A) 10 cm of pure toluene, (B) 1 cm of a solution of 0.08 g/liter M$_2$-BOBOB in toluene, and (C) 10 cm of a solution of 0.08 g/liter M$_2$-BOBOB in toluene (23).

length for the photons before they escape from the scintillator solution. At this time it should be mentioned that self-absorption problems can be reduced by the use of secondary solutes that act as wavelength shifters. This is discussed in the section on secondary solutes.

A solution of a scintillator solute was excited by a high-intensity light source. The photons emitted from the front surface of this solution were essentially undisturbed by any solute concentration effects. To study the effects of solute concentration and length of solution between photon emission and detection, Leggate and Owen made a simple experiment (23). The emitted photons were allowed to pass through (a) 10 cm of pure toluene, and (b) 1 cm and (c) 10 cm of an identical solution, respectively. The results are shown in Fig. III-9.

The choice of scintillator solute is also dependent on the spectral response of the detection system, i.e., the multiplier phototube. Ideally, with the new bialkali phototubes, a single scintillator solute such as PPO or PBD should give the maximum scintillation yield. The peak response of the phototube and the peak emission of these scintillator solutes occur at the wavelengths near 385 nm. However, in liquid scintillation counting one has to consider the total solution. There may still be valid reasons for adding a wavelength shifter. If there is a component of the scintillator solution which has a strong absorption band in the wavelength region of the emission of the primary solute, a secondary solute would shift the photon distribution to a region where the self-absorption is less. Therefore if the product of the flux of photons times the response of the phototube is increased, the result will be an increase in the scintillation yield. The real scintillation yield is given by the equation

$$S_x = \text{(number of photons of wavelength } \lambda)$$
$$\cdot \text{(response efficiency of the MPT for photons of wavelength } \lambda).$$

Consider a scintillator solution that produces 10,000 excited solute molecules. Let the solution have a component which has a strong absorption in the region of primary solute emission. A multiplier phototube is used which has a response of 28% for primary solute emission and only 10% for secondary solute emission. Consider the photon yields given in Table III-10.

Elimination of Concentration Quenching

Since the effects of interactions between solute molecules can lead to a decrease in the scintillation yield, studies have been made to determine which types of molecules are susceptible to concentration quenching (24–30). Most molecules which undergo concentration quenching have the ability for the chromophoric part of the molecule to obtain a planar configuration which is unhindered by the approach of a second like molecule. Take the case of anthracene (I). Anthracene has a planar chromophore

I

and undergoes very strong self-quenching at moderate concentrations of anthracene. Derivatives of anthracene with groups in the 9- and 9,10-positions have marked effects upon the self-quenching properties of the compounds. Methyl groups in the 9- and 9,10- positions do not affect the self-quenching as these groups are small and do not hinder the close approach of two anthracene molecules. However, large phenyl groups in the 9,10- positions will completely eliminate self-quenching. These properties are summarized in Table III-11.

Table III-10

Photon yields

Without secondary solute	With secondary solute
8400 photons	8400 photons
($\Phi = 0.84$)	(transfer = 100%, $\Phi = 0.84$)
2100 photons emitted	8400 photons
(75% absorption)	(no absorption)
588 relative response	840 relative response
(at 28%)	(at 10%)

Table III-11

Self-quenching and excimer fluorescence properties of anthracene and some 9- and 9,10-substituted derivatives

Substituent	Does self-quenching occur?	Has excimer fluorescence been measured?
None	Yes	? (Very little)
9-Methyl	Yes	Yes
9,10-Dimethyl	Yes	Yes
9,10-Diphenyl	No	No

Another example of the effects of molecular constitution on concentration quenching is illustrated by a series of derivatives of 2-phenyl indole (II) (28–30) where R and R′ are varied groups. Table III-12 lists the effects

II

of the lengths of a bridge between the 3- position of the indole and the 2′-position of the phenyl group.

A second technique of altering the structural constitution to eliminate self-quenching is called introduction of "bumper groups" (28). These groups are not a part of the chromophore group, but are large bulky groups that interfere with two molecules coming together in the proper orientation

Table III-12

**Structural effects of derivatives of 2-phenyl indole
on self-quenching and excimer formation**

Compound	Can this chromophore become planar?	Does this molecule self-quench?	Has excimer fluorescence been measured?
	Yes	Yes	No
	Yes	Yes	Solubility too low
	Yes	Yes	Yes
	No	No	No

for energy exchange or excimer formation. Consider the example of *p*-quaterphenyl (III) and some special derivatives of *p*-quaterphenyl. The

III

whole molecule forms the chromophore group as all four phenyl groups are conjugated. Table III-13 lists some observations of the scintillation yields of *p*-quaterphenyl and some of its derivatives. The substitutions of the very large bulky oxyalkyl groups at both ends of *p*-quaterphenyl has eliminated the self-quenching even at extremely high concentrations (~ 370 g/liter). The solubility limit of *p*-quaterphenyl of 0.12 g/liter does not allow for determination of the degree of self-quenching of this molecule.

The *p*-quaterphenyl chromophore planarity can be destroyed by the introduction of methyl groups on adjacent phenyl groups in positions such that the phenyl groups are forced out of a common plane. The tetramethyl derivative of *p*-quaterphenyl (Table III-13) has no self-quenching at any concentration, even at those approaching the pure state. Figures III-10 and III-11 show the scintillation yields as a function of concentration for the tetramethyl-*p*-quaterphenyl and two di-(oxyalkyl)-*p*-quaterphenyls.

Increased Solute Solubility

Many compounds which appear to have properties that would make them good scintillator solutes were often rejected because of their limited solubility in the normal scintillator solvents. The technique of introducing small alkyl groups onto these molecules has led to a marked increase in the solubilities with very little effect upon the fluorescence properties of the basic chromophore (31–33).

Many investigators did not use PBD because of its limited solubility, even though early investigations showed that for the same energy deposited in the scintillator solutions 12% more photons were emitted when the solute was PBD as compared to when it was PPO (1, 2, 33, 34). The synthesis of derivatives of PBD has greatly increased the solubility with essentially no change in the fluorescence properties. Table III-14 lists the properties of PBD and its derivatives.

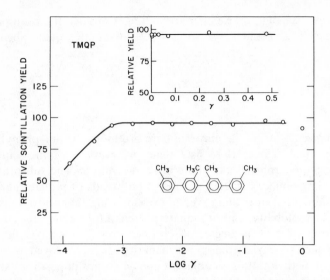

Fig. III-10.   Relative scintillation yield of tetramethyl-*p*-quaterphenyl as a function of concentration; γ is the ratio of moles of solute (n_2) to the total moles: $\gamma = n_2/(n_1 + n_2)$, where n_1 is moles of solvent.

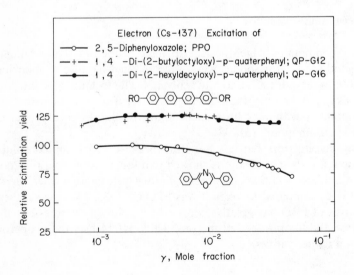

Fig. III-11.   Relative scintillation yields of two di-(oxyalkyl)-*p*-quaterphenyls as a function of concentration. Comparison of PPO is shown.

Table III-13

Elimination of self-quenching of *p*-quaterphenyl chromophore by "bumper effect" and "destroying" coplanar chromophore

	Solubility limit (g/liter)	Does self-quenching occur?
	0.12	? (Probably if solubility could be increased)
	370	No
	∞ (Forms a stable glassy-state)	No

Table III-14

Chemical and fluorescence properties of PBD and its derivatives

Structure	Abbreviation	Toluene solubility (g/liter, 20°C)	Fluorescence wavelength maximum (nm)	Decay time (10⁻⁹ sec)	Φ^a
	PBD	12	375	1.2	0.69
	i-Propyl-PBD	130	375	1.2	0.69
	t-Butyl-PBD	130	375	1.2	0.69
	p-Methyl-PBD	60	375	1.2	0.69
	o-Methyl-PBD	60	375	1.2	0.69

[a] Quantum yield as reported by J. B. Birks, "Photophysics of Aromatic Molecules." Wiley, New York, 1970.

The compound POPOP (IV) was used as a secondary solute for many

IV

years. The limited solubility and slow rate of dissolution of POPOP made the preparation of scintillator solutions time consuming. Prolonged gentle heating (often 24 hr) was necessary to dissolve the POPOP at the desired concentrations of 0.1 g/liter in toluene. The dimethyl derivative of POPOP (V) has a solubility of about 2 g/liter at room temperatures, almost four

(M_2-POPOP)

V

times more soluble than POPOP. But even more important, the desired amounts are readily soluble in toluene, and the fluorescence properties were only slightly altered. Table III-15 summarizes the data.

Table III-15

Comparison of properties of POPOP and the dimethyl derivative of POPOP, M_2-POPOP

Compound	Solubility (g/liter, 20°C)	Average fluorescence wavelength (nm)	Decay time (nsec)	Φ
POPOP	0.5	415	1.5	0.77
M_2-POPOP	2.0	427	1.5	0.77

The solubility of a basic type of compound can also be increased by altering the length and branching of the substituent group. Table III-16 shows the change of solubility of a series of *p*-quaterphenyls and *p*-quinque-phenyls (32).

Solute Stability

An important property of a solute which is often not considered is its stability. It should not react with various types of samples or agents used

Table III-16

Solubility change in *p*-quaterphenyls and *p*-quinquephenyls

R Group	Solubility (g/liter)	Solubility (g/liter)
H	0.12	0.005
Methyl	3.7	0.1
Ethyl	7.5	0.18
n-Butyl	43	0.48
n-Hexyl	46	0.52
n-Octyl	48	0.55

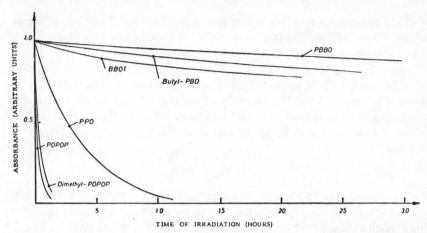

Fig. III-12. Photochemical degradation of some scintillators when irradiated by a 1500-W xenon lamp. Solvent, toluene; concentration, 0.5 g/liter; irradiation intensity, 180,000 lx.

in solubilizing samples, and it should also be stable toward radiation decomposition (33, 35). Figure III-12 shows the effects of irradiation on toluene solutions of several scintillator solutes with uv and visible light from a xenon lamp (33). The solutions were air saturated, which undoubtedly played a role in the degradation of some of the solutes.

Secondary Solutes

In the early studies of counting with liquid scintillator solutions it was known that addition of small concentrations of a second solute to the solution greatly increased the scintillation efficiency (36, 37). The concentrations of the secondary solute were often only 1% of the concentration of the primary solute. The reason for the increase in scintillation yield was due to a shift in the distribution of the energies of the emitted photons to a lower energy band (i.e., longer wavelengths). The new wavelength distribution produced a greater response from the multiplier phototubes.

These secondary solutes got the name of "wavelength shifters" as a result of their use in this manner. Since those early days the development of new multiplier phototubes with different wavelength responses, namely in the region of primary solute emission, has almost eliminated the need to use a secondary solute that functions as a wavelength shifter.

However, there are occasions when the use of a secondary solute is still desirable. If a scintillator solution has present in it a compound which has a strong photon absorption probability for the photons emitted by the primary solute, a secondary solute could produce a different distribution which would not be affected as much by the absorbing substance. Table III-17 lists some of the compounds that have been used as secondary solutes. Of these dimethyl-POPOP is the most commonly used secondary solute.

In all large-volume applications of liquid scintillators it is necessary to use a secondary solute. The intensity of the transmitted photons is related to the solute concentration c, the length of solution through which the photons have to travel l, the extinction coefficient ϵ, and the factor for overlap f:

$$I = fI_0 e^{-lc\epsilon}.$$

Thus if the overlap factor and the values of ϵ of the primary and secondary solutes are the same, the path length can be increased 100-fold, provided the concentration of the secondary solute is only 1% of the primary solute, and the light transmission will be equal to solution with no secondary solute.

An amazing property of secondary solutes is that the photons produced by the scintillator solution come only from the secondary solute, provided the concentration is sufficient to provide 100% energy transfer from the primary to the secondary solute (38–40). The energy transfer is essentially quantitative at secondary solute concentrations of 0.05–0.2 g/liter of most compounds used as secondary solutes.

The energy transfer from primary solute to the secondary solute does not occur by the emission of photons by the primary solute, absorption by the secondary solute, and reemission. This series of events would take considerably longer than the time of emission observed. The decay times of solutions with solvent, primary, and secondary solutes were measured by excitation of the solvent in one experiment and direct excitation of the secondary solute in a second experiment (41, 42). The decay time of the photon intensity was identical in both experiments. Thus the energy transfer processes were many times faster than the decay time of the secondary solute, so as not to alter the observed decay time. Since the decay time of the primary solute was comparable to that of the secondary solute, it was not possible that the emission–absorption process could have occurred.

Some Common Liquid Scintillators

There are an infinite number of liquid scintillator solutions that can be prepared which will be efficient. Many of these solutions will be tailored for a particular use or type of sample. In Table III-18 are listed a few examples of some commonly used scintillator solutions.

Table III-17

Some commonly used secondary solutes

Compound	Abbreviation	Concentration (g/liter)	Average wavelength (nm)
	POPOP	0.05–0.2	415
	M₂-POPOP	0.1–0.5	427
	α-NPO	0.05–0.2	400
	bis-MSB	1.5	425

Table III-18

Some commonly used liquid scintillator counting solutions

Solvent	Primary solute (g/liter)	Secondary solute (mg/liter)	Other additives	Type of sample
Toluene	PPO (4–6)	M₂-POPOP (50–200)	—	Organic soluble
Toluene	Butyl-PBD (8–12)	—	—	Organic soluble
p-Xylene	Butyl-PBD (8–12)	—	Ethanol	Aqueous
Dioxane (Bray's solution)	PPO (4)	M₂-POPOP (200)	Methanol (100 ml) Ethylene glycol (200 ml) Naphthalene (60 g)	Aqueous
Dioxane	PPO (7)	M₂-POPOP (300)	Naphthalene (100 g)	Aqueous
p-Xylene	PPO (4–6)	—	Emulsifier (i.e., Triton N-101)	Water and aqueous

The first two solutions need little comment. They are used mainly for counting organic soluble-type samples. Without further additives these solutions usually constitute the choice for homogeneous solution counting. The use of a wave shifter (M_2-POPOP) will be determined by the type of multiplier phototube and the need to reduce effects of certain color quenchers. The use of butyl-PBD over PPO is recommended where the maximum scintillation yield is required. Butyl-PBD has about a 12% greater scintillation yield than PPO in air-saturated scintillator solutions at optimum solute concentrations: 8–12 g/liter of butyl-PBD and 4–6 g/liter of PPO.

The last four solutions listed are used for counting aqueous solutions. Ethanol can be used to incorporate small amounts of aqueous solutions with still quite high scintillation efficiencies. The stability of dioxane and chemiluminescence problems have placed the use of dioxane in some disfavor. The use of aromatic solvents (e.g., *p*-xylene) and emulsifiers (e.g., Triton X-100) has shown great potential for counting aqueous solutions. However, phase changes in the solution resulting from variations of aqueous concentrations do occur. These solutions are not homogeneous, and normal quench correction methods cannot always be used for determination of the sample counting efficiency.

References

1. F. N. Hayes, B. S. Rogers, R. Sanders, R. L. Schuch, and D. L. Williams, Rcp. LA-1639. Los Alamos Sci. Lab., Los Alamos, New Mexico, 1953.
2. F. N. Hayes, B. S. Rogers, and P. C. Sanders, *Nucleonics* **13** (1), 46 (1955).
3. C. W. Lawson, F. Hirayama, and S. Lipsky, *J. Chem. Phys.* **51**, 1590 (1969).
4. P. Skarstad, R. Ma, and S. Lipsky, *in* "Organic Scintillators" (D. L. Horrocks, ed.), p. 3. Gordon & Breach, New York, 1968.
5. D. L. Horrocks, *J. Chem. Phys.* **52**, 1566 (1970).
6. R. Cooper and J. K. Thomas, *J. Chem. Phys.* **48**, 5097 (1968).
7. J. K. Thomas, Private communication, 1969.
8. J. B. Birks, *in* "Organic Scintillators and Liquid Scintillation Counting" (D. L. Horrocks and C. T. Peng, eds.), p. 3. Academic Press, New York, 1971.
9. A. Skerbele and E. N. Lassettre, *J. Chem. Phys.* **42**, 395 (1965).
10. H. B. Klevens and J. R. Platt, *J. Chem. Phys.* **17**, 470 (1949).
11. G. Laustriat, R. Voltz, and J. Klein, *in* "The Current Status of Liquid Scintillation Counting" (E. D. Bransome, Jr., ed.), p. 13. Grune & Stratton, New York, 1970.
12. G. K. Oster and H. P. Kallmann, *Nature (London)* **194**, 1033 (1962).
13. G. K. Oster and H. P. Kallmann, *Int. Symp. Luminescence* (N. Riehl, H. Kallmann, and H. Vogel, eds.) p. 31. Thiemig, München, 1966.
14. G. K. Oster and H. P. Kallmann, *J. Chem. Phys.* **64**, 28 (1967).
15. R. Voltz, *Radiat. Res. Rev.* **1**, 301 (1968).
16. A. Kearvell and F. Wilkinson, *in* "Organic Scintillators" (D. L. Horrocks, ed.), p. 69. Gordon & Breach, New York, 1968.
17. I. B. Berlman, *J. Chem. Phys.* **33**, 1124 (1960).

18. Th. Förster and K. Kasper, *Z. Elektrochem.* **59**, 977 (1955).
19. B. Stevens, *Nature (London)* **192**, 725 (1961).
20. J. B. Birks, *Nature (London)* **214**, 1187 (1967).
21. D. L. Horrocks, *in* "Organic Scintillators and Liquid Scintillation Counting" (D. L. Horrocks and C. T. Peng, eds.), p. 75. Academic Press, New York, 1971.
22. I. B. Berlman, "Handbook of Fluorescence Spectra of Aromatic Molecules." Academic Press, New York, 1965.
23. P. Leggate and D. Owen, *in* "Organic Scintillators" (D. L. Horrocks, ed.), p. 357. Gordon & Breach, New York, 1968.
24. M. Furst and H. Kallman, *Phys. Rev.* **109**, 646 (1958).
25. M. Furst and H. Kallmann, *in* "Liquid Scintillation Counting" (C. G. Bell and F. N. Hayes, ed.), p. 237. Pergamon, Oxford, 1958.
26. J. B. Birks and L. G. Christophoron, *Nature (London)* **194**, 442 (1962).
27. B. Stevens, *Spectrochim. Acta* **18**, 439 (1962).
28. D. L. Horrocks and H. O. Wirth, *in* "Organic Scintillators" (D. L. Horrocks, ed.), p. 375. Gordon & Breach, New York, 1968.
29. D. L. Horrocks and H. O. Wirth, *J. Chem. Phys.* **49**, 2907 (1968).
30. D. L. Horrocks, *J. Chem. Phys.* **49**, 2913 (1968).
31. H. O. Wirth, *Proc. Conf. Organic Scintillation Detectors, Univ. of New Mexico, 1960* (G. H. Daub, F. N. Hayes, and E. Sullivan, eds.), Rep. TID-7612, p. 78. U. S. At. Energy Commission, Office, Washington, D. C., 1961.
32. H. O. Wirth, F. U. Herrmann, G. Herrmann, and W. Kern, *in* "Organic Scintillators" (D. L. Horrocks, ed.), p. 321. Gordon & Breach, New York, 1968.
33. E. Kowalski, R. Anliker, and K. Schmid, *in* "Organic Scintillators" (D. L. Horrocks, ed.), p. 403. Gordon & Breach, New York, 1968.
34. B. Scales, *Int. J. Appl. Radiat. Isotop.* **18**, 1 (1967).
35. M. E. Ackerman, G. H. Daub, F. N. Hayes, and H. A. Mackay, *in* "Organic Scintillators and Liquid Scintillation Counting" (D. L. Horrocks and C. T. Peng, eds.), p. 315. Academic Press, New York, 1971.
36. H. Kallmann and M. Furst, *Nucleonics* **8** (3), 32 (1951).
37. F. N. Hayes, D. G. Ott, and V. N. Kerr, *Nucleonics* **14** (1), 42 (1956).
38. I. B. Berlman, *J. Chem. Phys.* **34**, 598 (1961).
39. D. L. Horrocks, *in* "Organic Scintillators" (D. L. Horrocks, ed.), p. 45. Gordon & Breach, New York, 1968.
40. H. Kallmann and M. Furst, *in* "Liquid Scintillation Counting" (C. G. Bell and F. N. Hayes, eds.), p. 3. Pergamon, Oxford, 1958.
41. W. R. Ware, *J. Amer. Chem. Soc.* **83**, 4374 (1961).
42. J. B. Birks and K. N. Kuchela, *Proc. Phys. Soc. (London)* **77**, 1083 (1961).

CHAPTER IV

LIQUID SCINTILLATION COUNTERS AND MULTIPLIER PHOTOTUBES

Almost all commercially available liquid scintillation counters are coincidence systems, that is, two multiplier phototubes viewing the same sample, although one manufacturer does offer a single phototube instrument (Nuclear Enterprises, Ltd.). The main advantage of the coincidence system is the reduction of background rates due to randomly generated pulses produced within each multiplier phototube, commonly called noise.

One main difference between commercial instruments has to do with the type of amplifications, of which there are two: linear and logarithmic. Linear amplification generates a pulse which is proportional to the output of the multiplier phototube. Since the output of the multiplier phototube is proportional to the energy deposited in the scintillator solution, at least for electrons, the linear amplification generates pulses proportional to the energy of the electron (beta particle) that produced the scintillation. Most instruments which have linear amplification have a separate amplifier for each counting channel. For instruments with three data channels there are three different amplifiers which can be set at the same or different gains to cover different energy ranges.

The logarithmic amplification system converts the summed output pulse of the two multiplier phototubes into a new pulse with an amplitude equal to the logarithm of the summed multiplier phototube pulses. By so doing it is possible to measure a wide range of pulse amplitudes with a single amplifier. The pulse amplitudes produced by electrons from zero energy to greater than the ^{32}P endpoint (1.7 MeV) can be handled in a single amplifier system due to the logarithmic conversion.

Figures IV-1 and VI-2 show the pulse distributions for instruments with linear pulse response and three amplifiers, and with logarithmic pulse response and a single amplifier, respectively. Both types of amplification give comparable results for counting efficiencies, background, and count rate of higher energy nuclides in the channel for counting lower energy nuclides (spillover). The same pulses are analyzed in both cases.

Block diagrams of instruments with the two types of amplification are given in Fig. IV-3.

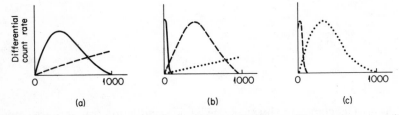

Fig. IV-1. Pulse distributions in linear amplifier-type counter: —^{3}H spectrum; — — —^{14}C spectrum; . . . . ^{32}P spectrum. Gains are (a) 100, (b) 10, and (c) 1.

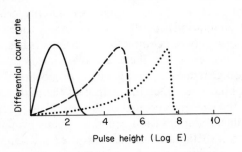

Fig. IV-2. Pulse distributions in logarithmic amplifier-type counter: —^{3}H; — — — ^{14}C; . . . . ^{32}P.

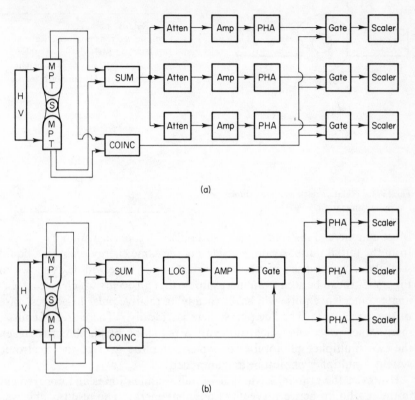

(a)

(b)

Fig. IV-3. Simplified block diagrams for (a) linear and (b) logarithmic amplifier–type counters.

Pulse Summation

The early models of liquid scintillation counters utilized the output of only one multiplier phototube for pulse height discrimination (1–3). The other multiplier phototube served as a coincidence monitor only. Thus the response to a given energy event could vary, depending on how the scintillation photons were divided between the two multiplier phototubes. Figure IV-4 shows a representation of the variation of pulse height due to the same total energy deposited in a scintillator solution. Of course the probability that the photoelectron production is 7–0 or 0–7 is very unlikely. But even with the most probable split of 3–4 and 4–3, the response will be spread over a relative pulse height distribution.

The pulse summation circuit adds together all pulses from the output of the multiplier phototube which occur within a certain time period.

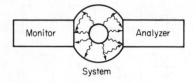

Relative number of photoelectrons produced in MPT		Relative pulse height response
Analyzer	Monitor	
7	0	0
6	1	6
5	2	5
4	3	4
3	4	3
2	5	2
1	7	0
0	7	0

Fig. IV-4. Relative response in instrument without pulse summation.

Most commercial instruments will add pulses that occur within 20–30 nsec from the initial pulse. Again, considering the same energy event in a liquid scintillator with pulse summation the relative response will be as shown in Fig. IV-5. It can be seen from this example that all events which give a legitimate coincidence pulse will have "ideally" a relative pulse height response of the same value. The key phrase here is "ideally." There are, of course, many other processes which can cause a reduction of photons that reach the two multiplier phototubes or which alter the gain of the electronic system—multiplier phototube and amplifier.

However, the primary advantage of pulse summation is an improvement in the resolution between events of different energy. The result of this was initially evident in a better separation between ^{3}H and ^{14}C pulses. Without

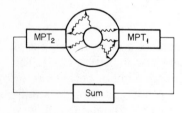

Relative number of photoelectrons produced in MPT		Relative pulse height response
No. 1	No. 2	
7	0	0
6	1	7
5	2	7
4	3	7
3	4	7
2	5	7
1	6	7
0	7	0

Fig. IV-5. Relative response in instrument with pulse summation.

pulse summation the ^3H and ^{14}C spectra gave distributions as shown in Fig. IV-6. In newer instruments with pulse summation it is now possible to detect about 70% of ^{14}C pulses above the ^3H endpoint (Fig. IV-7).

Another advantage of pulse summation is the increased response of coincidence events over noncoincidence noise, i.e., an increase in the signal-to-noise ratio. This would not lead to an increase in coincidence events, but could lead to small increases in counting efficiency by increasing the pulse height of some coincidence events above the noise threshold. Consider a series of pulses produced with various pulse heights in two multiplier phototubes as shown in Fig. IV-8.

Without pulse summation two pulses of those shown would be analyzed and registered in one of the scalers, provided the discriminators were set to accept those pulses. In most events shown either the analyzer pulse or the monitor pulse failed to exceed the threshold level of the pulse height analyzer. With the pulse summations the same two pulses were analyzed and registered in the scalers. Each pulse amplitude exceeds the threshold by a greater margin and the amplitudes are more nearly equal.

The amplitude of one of the accepted pulses without pulse summation was only slightly above the threshold level. With pulse summation that pulse amplitude increased almost fivefold above the threshold. However, it should be remembered that pulse summation can lead to only a small increase in counting efficiency for very low-energy beta emitters, i.e., ^3H, and usually no increase for higher energy beta emitters, i.e., ^{14}C.

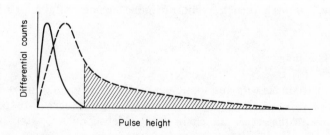

Fig. IV-6. Relative pulse distributions for ^3H and ^{14}C in instrument without pulse summation. Shaded area represents 40% of ^{14}C pulses above ^3H endpoint.

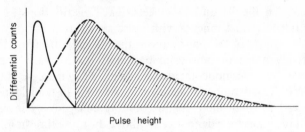

Fig. IV-7.   Relative pulse distributions for ^{3}H and ^{14}C in instrument with pulse summation. Shaded area shows 70% of ^{14}C pulses above ^{3}H endpoint.

Mode of Standardization

Most commercial instruments provide external standardization by a γ-ray source (4, 5). Some instruments give the external standardization by counts or counts per minute (cpm) in a given window. Other instruments will count the external standard source in two channels for calculation of the channels ratio.

Instruments will vary in data printout. Some give gross counts only, to which the operator has to apply corrections for sample counts in the counting channel or channels, and calculate the ratio also. Instrument designs vary from this simple format all the way to those which have automatic correction for sample contribution to external standard counting data, two separate counting channels for external standard, and automatic calculation and printout of the external standard channels ratio.

Also many commercially available instruments have the capability of calculating the ratio of sample counts (or cpm) in two channels. In this mode of data presentation, the proper choice of channels will allow for use of the ratio as a quench monitor (sample channels ratio method) (6).

Sample Handling

Most instruments with automatic sample handling of 50 or more samples use the serpentine chain sample conveyors. Instruments are presently available with the serpentine chain drive which handle 50, 100, 200, and 300 samples. However, there are some instruments available which use the tray concept of sample handling. Each tray is loaded with the desired samples and the tray is moved into position. The samples are usually removed from the tray, by row, to the elevator position for introduction into the counting well. There are also some instruments in which the samples are manually introduced one at a time.

The samples are usually lowered from the sample storage area into the counting region by an elevator mechanism. The elevator system is made light-tight with shutters, O-rings, and felt pads to prevent ambient light excitation of the multiplier phototubes during the descent and ascent of the sample. The applied voltage can remain on at all times during normal operation because of the light-tight elevator mechanism.

Correction for Sample Quench

Two commercial manufacturers offer methods which will automatically correct sample counting conditions to correct for variations in sample quench. Packard Instruments offers the Absolute Activity Analyzer, Packard Model 544AAA. Beckman Instruments' quench correction system is called Beckman AQC, Automatic Quench Compensation. In both systems the gain of the system is altered to satisfy predetermined counting conditions.

The Packard system utilizes a magnetic defocusing coil which reduces the gain of the multiplier phototube until the external standard channels

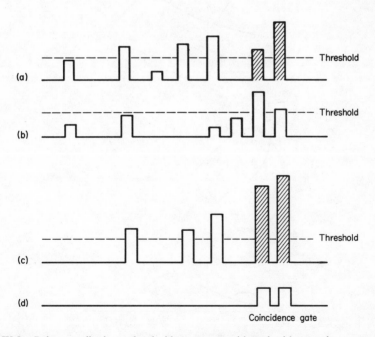

Fig. IV-8. Pulses amplitudes and coincidence events with and without pulse summation, where the shaded bars are the registered counts. (a) Pulses from MPT_1 (analyzer), (b) pulses from MPT_2 (monitor), (c) summation of pulses from MPT_1 and MPT_2 greater than threshold, (d) other coincidence events lost due to finite threshold in coincidence gate circuit.

ratio value is identical with a given value for which the counting efficiency
has been previously determined. The sample count rate is divided by the
efficiency value, which is dialed on a series of switches, and the data printed
out as sample disintegrations per minute (dpm).

The Beckman system operation increases the system gain by increasing
the multiplier phototube applied voltage. The external standard channels
ratio of each sample is measured at the base high voltage. The system gain
is increased a certain amount which has been previously determined by a
set of quenched standards. The gain restoration is chosen such that the
endpoint energies of ^3H and ^{14}C are restored to the same discriminator
settings as for the sample with the least quench of the standard quenched
series. Figure IV-9 illustrates the effect of the gain increase.

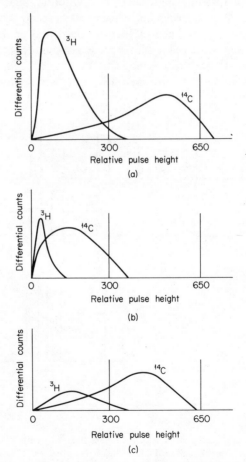

Fig. IV-9.   The effect of gain restoration on the relative pulse height distributions of ^3H and
^{14}C. (a) Unquenched; (b) quenched; (c) quenched with gain restored.

One difference between these two systems involves the counting efficiency after the alteration of the system gain. In the Packard system the counting efficiency is reduced to meet the predetermined values. Thus counts are actually discarded, which could be very critical with low count rate samples. In the Beckman system the counting efficiency remains the same, and in some cases shows a slight increase.

Another difference revolves around the measurement of dual-labeled samples. In the Packard system, with fixed counting channels, the number of counts of the higher energy isotope (^{14}C) which occur in the lower energy isotope (^{3}H) counting channel will increase. However, in the Beckman system, in which the endpoint energies are restored to the proper discriminator settings, the number of counts of the higher energy isotope (^{14}C) which occur in the counting channel for the lower energy isotope (^{3}H) are actually reduced. Under normal operations Beckman AQC will give a constant counting efficiency of ^{14}C in the ^{3}H counting channel, which greatly simplifies correction of counting data of dual-labeled samples.

Data Computation

Some of the instruments have computational capabilities built into the instrument to calculate such data as counts per minute, error function (σ or 2σ), sample channels ratio, and external standard channels ratio. Also there are some instruments which are now available with dedicated computers.

Multiplier Phototubes*

The multiplier phototubes are a very important link in the chain of components that make up the liquid scintillation counter. If the efficiency of the multiplier phototube is low, the lost events (counts or pulses) can never be regained no matter how fine the rest of the electronics or how unquenched the sample preparations.

There are at least three main parts of the multiplier phototube which are critical in determination of its efficiency and response. These are:

(1) the photocathode, which converts the photons into a corresponding number of electrons defined by the wavelength (energy) of the photons, number of photons, the quantum efficiency of the photocathode, and the composition of the photocathode;

* For more detailed information, see manufacturers' manuals (7).

(2) the dynodes, which multiply the number of electrons by the process of converting one imparting electron into three or four departing electrons;

(3) the anode, which collects electrons from the dynodes and converts the total charge into a pulse of amplitude directly proportional to the number of electrons collected.

The Photocathode

The face of the multiplier tube is made of glass. When photons of energy in the ultraviolet region are to be measured the face is made of quartz, which is highly transmissive in that energy range (see Fig. IV-10). The photocathode is deposited on the inside of the face and is composed of a special combination of materials that are volatilized onto the glass to form a thin layer which is capable of absorbing the photon energy and as a result releases electrons. These electrons are called photoelectrons. The photoelectrons are accelerated and focused by potential differences between the photocathode and first dynode and the field created by the focusing electrodes. Figure IV-11 shows the path of the photoelectrons. Photoelectrons created at the edges of the multiplier phototube are accelerated and their direction of flight altered by the field created by the interactions of the potentials on the focusing electrodes. By proper selection of the potentials all the photoelectrons are focused on the first dynode.

The photocathode efficiency is not only a function of the wavelength of the photons but it will also vary over the face of the photocathode. Some phototube efficiencies vary as much as tenfold over the area of the photocathode. Figure IV-12 shows a map of the relative efficiency of a photocathode for monochromatic photons. There is often a spot of high quantum efficiency. The variation of efficiency is likely due to a combination of focusing and nonuniform deposition of the photocathode material.

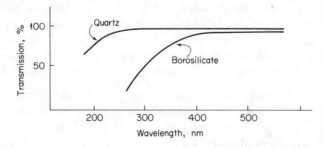

Fig. IV-10.   Transmission properties of quartz and borosilicate (i.e., Pyrex).

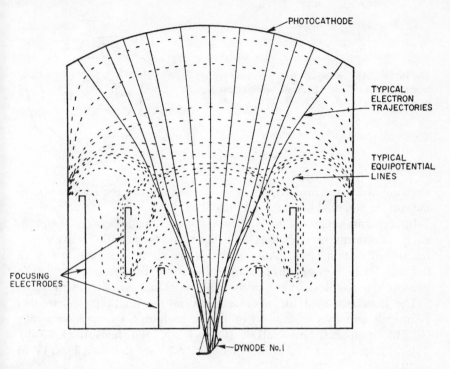

Fig. IV-11. Diagram showing potentials causing acceleration and focusing of photoelectrons on first dynode.

The interaction of the photons with the photocathode material is a process called *photoemission*, the end result of which is the liberation of electrons. The process can be considered in three separate steps:

(a) absorption of the photon and transfer of its energy to an electron in the photocathode material,

(b) migration of the electron to the surface (toward the inside surface), and

(c) escape of the electron into the vacuum (overcoming the potential barrier).

Energy losses, with resultant efficiency decreases, occur in each of these processes. Not all photons are absorbed, many are lost by reflection and transmission. Also some scattering events may or may not result in electron emission due to only part of the photon energy being transferred to the electron. The photoelectrons can lose energy as the result of collisions with other electrons or with atoms of the photocathode material. Finally, some

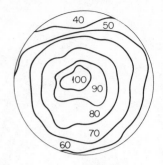

Fig. IV-12. Plot of equal response areas of a typical photocathode. The most sensitive area is given a value of 100.

electrons will have insufficient energy to escape from the surface due to the surface potential barrier (work function).

In preparing a photocathode, a compromise is made between the probability of absorption of photons (thick layer) and the probability of electron transmission (thin layer). Also materials are chosen to have low work function for high escape probability, but a work function high enough to prevent appreciable amounts of thermal (spontaneous) electron emission.

The materials used for photocathodes in the multiplier phototubes commonly employed in liquid scintillation counters are cesium–antimony (Cs_3Sb), multialkali or trialkali ($CsNa_2KSb$), and bialkali (K_2CsSb). Typical response curves for these photocathodes are shown in Figs. IV-13 and IV-14.

Another important factor which affects the response of the photocathode is the glass face of the multiplier phototube. Use of quartz will allow the measurement of more of the ultraviolet radiation that is normally absorbed in borosilicate-type glasses. An example of the effect of SiO_2 is shown in the wavelength of maximum response (λ_{max}) and quantum efficiency (q) at λ_{max} for a Cs_3Sb photocathode:

$$\text{with quartz:} \quad \lambda_{max} = 330, \quad q = 24\%,$$

$$\text{with borosilicate:} \quad \lambda_{max} = 400, \quad q = 13\%.$$

Not only is the λ_{max} shifted to lower wavelengths but q increases, which indicates that an increased number of photons are passed by the quartz face.

Dynodes

There are basically two types of dynode structures: One is called the venetian blind (Fig. IV-15) and the other the linear (Fig. IV-16).

The dynode structure is the multiplication part of the multiplier phototube. The photoelectrons are accelerated and focused on the first dynode.

When these high-energy electrons strike the dynode, a number of secondary electrons are emitted from the dynode. The number of secondary electrons emitted per photoelectron will be a function of the kinetic energy of the photoelectrons and the work function of the dynode material.

Again the processes which occur in the yield of secondary electrons are very similar to the processes in the production of the photoelectrons at the photocathode except that electrons rather than photons strike the dynodes. Essentially three processes occur:

(a) high-energy electrons excite electrons of the dynode material,
(b) the excited electrons migrate toward the surface, and
(c) electrons with energy greater than the surface barrier will escape from the surface.

Those that escape are accelerated by a potential difference to the next dynode where the same processes occur again.

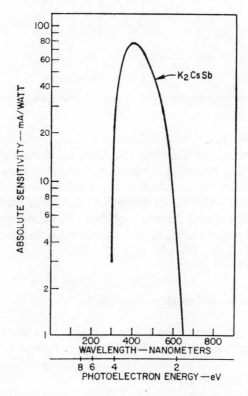

Fig. IV-13. Spectral response of typical bialkali photocathode.

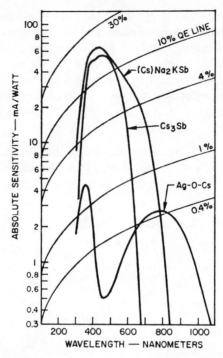

Fig. IV-14. Spectral response of several types of material used as photocathodes.

The gain of the multiplier phototube is defined as the factor relating to the increase in electrons collected at the anode for a single electron striking the first dynode. The liberation of secondary electrons is subject to statistical fluctuations. On the average about four secondary electrons are liberated at each dynode for each electron impinging on the dynode. This figure can vary. Considering only the Poisson probability function for small numbers, if the average expected yield is four, the occurrences and their probabilities are those given in Table IV-1.

Number of Electrons per Incident Electron

There is a probability that less than and greater than the expected average number of electrons will be ejected. However, if the average number of electrons (four) is ejected at each of 10 dynodes, the gain for a single electron striking the first dynode will be

$$\text{gain} = 4^{10} \approx 10^6,$$

Table IV-1

Poisson probability for expectation of four electrons

Number of electrons	Fraction of events that give listed number of electrons
0	0.018
1	0.073
2	0.147
3	0.195
4	0.195
5	0.156
6	0.104
7	0.059
8	0.030
9	0.013
10	0.005
11	0.002
≥ 12	0.003
Total	1.000

a gain of one million. If the average were only 3.5 electrons per incident electron, the gain would be 3.5^{10} or 2.6×10^5, a factor of almost 4 less.

If the mathematical calculations are considered, it will become readily evident that the statistical fluctuations of the first and second dynode will determine the statistical fluctuations of the final gain. This is because after the first two dynodes the number of electrons is sufficiently large that the fluctuations of any single electron will not be large enough to affect the final gain.

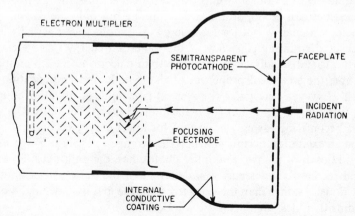

Fig. IV-15. Schematic diagram of venetian-blind type of multiplier structure.

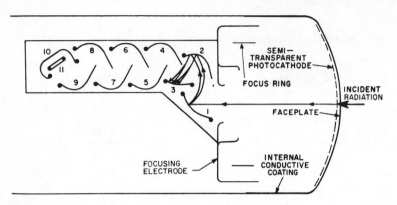

Fig. IV-16.   Schematic diagram of linear type of multiplier structure: 1–10, dynodes-electron multiplier; 11, anode.

The number of secondary electrons per incident electron is partly dependent on the kinetic energy of the incident electron. Thus increasing the high voltage of the multiplier phototube, which increases the potential difference between dynodes, will increase the kinetic energy that the electrons obtain before they strike the following dynode. Thus the gain is usually increased with an increase in the high voltage.

The type of dynode construction is important in determination of the pulse transit time. The venetian blind structure has a relatively slow time response, while the linear multiplier structure has fast time response. The time response of both types of tubes is adequate for counting rates normally used in liquid scintillation counting ($\leq 10^6$ cpm). However, the fast transit times are important for pulse shape discrimination.

In liquid scintillation counting the statistical nature and amplification characteristics can be important in determining not only the response per unit excitation energy, but for low-energy excitations it can actually affect the measured counting efficiency. If the excitation energy is large enough to produce a statistically large number of photoelectrons in each multiplier phototube, the statistical properties of the dynode and amplification will be small. However, since most commercial-type liquid scintillation counters have a small input pulse bias (to reduce noise pulses), the statistical probabilities become increasingly important for lower energy excitations.

When an excitation occurs which on the average will produce one photoelectron in each of the two multiplier phototubes, the multiplication in the first and second dynodes may well determine if the pulse will be counted. If the gain is 3, rather than the expected 4, on the first dynode, and 4 on all succeeding dynodes, the final pulse will be

$$3 \times 9^4 = 7.8 \times 10^5 \text{ electrons.}$$

This is 22% less than the 10^6 electrons for four electrons per dynode. If the gain was 3 for both the first and second dynodes, followed by 4 for all the remaining dynodes, the final pulse will be

$$3 \times 3 \times 8^4 = 5.9 \times 10^5 \text{ electrons.}$$

This is 41% less than the 10^6 electrons normally expected (Fig. IV-17). If the threshold limit is set to accept just single photoelectrons which get the total gain of 10^6, both of these pulses would be rejected. This will be reflected in a loss of counts.

It can be seen by the simplified calculations shown in Table IV-2 that once past the first or second dynodes, statistical variations will have little effect on the total gain. If one of the electrons has a gain of 3, rather than the expected 4, there will be increasingly less effect on the total gain the further down the dynode string the gain variation occurs.

Anode

The anode acts as a collector for the electrons leaving the last dynode; it collects all or a part of the electrons. The resulting collection produces a voltage which is converted into a voltage pulse through a capacitor. The amplitude of the voltage pulse is proportional to the number of electrons collected on the anode, which in turn are proportional to the number of photoelectrons, which are proportional to the number of photons, which are proportional to the energy of excitation particle.

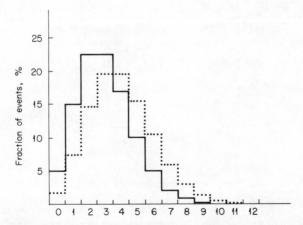

Fig. IV-17. Variation of distribution of electrons produced for expectation of three (—) and four (....) electrons.

Table IV-2

Effect of variation in secondary electron production on total gain as function of dynode on which variation occurs

Dynode 1	Dynode 2	Dynode 3	Dynodes 4–10	Total gain
4	4	4	4	1.05×10^6
3	4	4	4	$7.8 \ \times 10^5$
4	$3 + (3 \times 4)^a$	4	4	$9.8 \ \times 10^5$
4	4	$3 + (15 \times 4)^a$	4	1.02×10^6

[a] One of the impinging electrons produced three secondary electrons, the rest produced four secondary electrons.

The response of a multiplier phototube is limited by the anode current capacity. Most multiplier phototubes are limited to a few milliamperes under normal operations. If the number of electrons becomes too great, a space charge will build up between the last dynode and the anode. This leads to a nonlinear response at high anode currents. Also, continued operation of a multiplier phototube under conditions of high anode current can lead to fatigue which will result in decreased gain.

Two counting conditions can lead to anode saturation or fatigue. These are high count rates and/or high-energy particles. Most multiplier phototubes only show serious effects when both of these occur at once, i.e., high count rates of high-energy particles such as ^{32}P beta particles. The multiplier

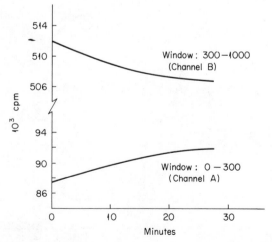

Fig. IV-18. Gain decrease as shown by relative counts in two channels with high count rate source of $^{14}C-^{14}C$-toluene, 1.6×10^6 dpm.

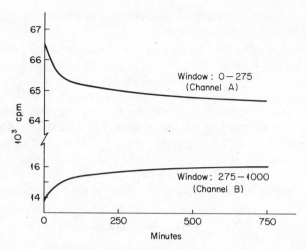

Fig. IV-19. Slow recovery of gain after conditions in Fig. IV-18 with moderate rate ^{3}H source—^{3}H-toluene, 1.26×10^5 dpm.

phototube will usually recover after anode saturation or dynode fatigue. Depending on the individual tube, the time for recovery can be from a few minutes to days.

Figures IV-18 and IV-19 show the results of a multiplier phototube becoming fatigued and its subsequent recovery. These data are not typical of all multiplier phototubes, but are shown as an example of the effects of fatigue on counting efficiencies and pulse height distributions. The counting channels A and B are set to count ^{3}H and ^{14}C above the ^{3}H endpoint (^{14}C/^{3}H), respectively. The data in Fig. IV-18 were obtained by counting a ^{14}C-containing sample of high count—1.6×10^6 dpm. It can be seen that the gain shift was very rapid. The count rate in channel A increased approximately 5% in 25 min, corresponding to a loss of counts in channel B. The number of counts lost from channel B is equal to the increase in the number of counts in channel A. Therefore, no counts were lost, i.e., the counting efficiency remained constant. However, the change in relative distribution of pulses was identical to a quenching process. Thus gain shifts can sometimes be confused with quenched samples.

After the ^{14}C sample was counted and the multiplier phototube showed a gain shift, a second sample of moderate count rate of low-energy events was placed into the counting chamber. The sample contained 1.26×10^5 dpm of ^{3}H and the counting efficiency was 63.9% in a wide open window, 51.2% in channel A, and 12.7% in channel B. It can be seen from Fig. IV-19 that there is a rapid change in the observed count rates in channels A and B with time. However, contrary to the ^{14}C results, there is a gain increase with time as evident by increasing count rate in channel B accompanied by the equivalent decrease in channel A.

If moderate count rates of samples are used, there will usually be no multiplier phototube gain shifts. However, when changing from a condition of low counting rate samples to high counting rate (or vice versa) it may be necessary to allow the multiplier phototubes to reach their new equilibrium gain. Often it is desirable to use the external standard γ-ray source to accelerate the gain change if samples of high count rates of high-energy particles are to be measured. The γ-ray will produce high-energy electrons in the scintillator solution, and most commercial instruments have γ-ray sources that will produce count rates of several hundred thousand counts per minute in an unquenched sample. If the high voltage has been off the multiplier phototubes, use of the γ-ray source will accelerate the gain shift to its equilibrium value after the high voltage is turned on.

Other Factors Affecting Multiplier Phototubes

Several environmental factors can affect the operation of multiplier phototubes.

Temperature

With most multiplier phototubes temperature changes are most often associated with a change in the dark current (noise). The noise increases with increasing temperatures. At very high temperatures it is possible to damage a multiplier phototube permanently by the large dynode currents resulting from high noise rates. It is usually recommended to operate at or below room temperatures. With the new multiplier photocathodes the dark current is such that operation at room temperature is possible with coincidence-type liquid scintillation counters.

Small changes are produced in the spectral response characteristics with temperature changes. Usually in the wavelength region of interest in liquid scintillation counting (300–500 nm) there is a slight negative response function. The response will decrease a fraction of a percent per increase in temperature of $1°C$. This is primarily due to an increase in the resistance of the photocathode material with a decrease in temperature.

Magnetic and Electrostatic Fields

External magnetic and electrostatic fields cause changes in the gain of the multiplier phototube by causing deflection of the electrons from their normal pathway in the multiplication and collection system. The gain change can be either an increase or decrease in final pulse amplitude, depending on the nature of the effects on the fields present in the normal operation of the multiplier phototube. Of a serious nature are factors which can introduce permanent magnetic fields in the tube or its surroundings. The long path of the photoelectrons from the photocathode to the first dynode makes this part of the tube very susceptible to the external magnetic and electrostatic fields.

In liquid scintillation counters the multiplier phototubes are operable at relatively high voltages (1800–2200 V). However, this potential is divided over the many dynodes (usually 10) so that there is much less potential between each step of the multiplication process. Thus even small external magnetic and electrostatic fields can affect the gain depending to a great extent on how and where they interact with the multiplication system. Most commercial liquid scintillation counters are supplied with mu-metal shields for the multiplier phototubes which reduce the effect of external magnetic fields. Any such shields should be maintained at the same potential as the photocathode. For most commercial liquid scintillation counters this means that the mu-metal shield should be grounded.

Others

Excessive vibration or shock can alter the proper operation of multiplier phototubes. They can, in some cases, actually cause physical damage to the components of the tube, causing shorts between the elements. Also rupture of the metal–glass seals at the pin positions can cause loss of vacuum in the tube. Shifting of the relative position of the elements can cause gain shifts by affecting the focusing and multiplication of the electrons.

Operation under conditions of high pressure (atmosphere or under water) should be undertaken with previous testing to be sure that the envelope of the tube will be able to withstand the added strain.

References

1. R. D. Hiebert and R. J. Watts, *Nucleonics* **11,** No. 12, 38 (1953).
2. R. D. Hiebert, *in* "Liquid Scintillation Counting" (C. G. Bell and F. N. Hayes, eds.), p. 41. Pergamon, Oxford, 1958.
3. L. E. Packard, *in* "Liquid Scintillation Counting" (C. G. Bell and F. N. Hayes, eds.), p. 50. Pergamon, Oxford, 1958.
4. T. Higashimura, O. Yamada, N. Nohara, and T. Shidei, *Int. J. Appl. Radiat. Isotop.* **13,** 308 (1962).
5. D. G. Fleishman and V. U. Glazunov, *Instrum. Exp. Tech.* (*USSR*) p. 472 (1962); *Prib. Tekh. Eksp.* **3,** 55 (1962).
6. L. A. Baillie, *Int. J. Appl. Radiat. Isotop.* **8,** 1 (1960).
7. For example, see, "RCA Photomultiplier Manual." RCA, Electron. Components, Harrison, New Jersey.

CHAPTER V

PARTICLE COUNTING TECHNIQUES

Liquid scintillators can and have been used to measure almost every kind of radioactive decay: negatron, positron, electron captive, gamma, alpha, fission, proton, neutron, neutrino, cosmic rays, etc. Some of these are discussed in this chapter, with emphasis on measurements of nuclides other than those commonly measured in liquid scintillation solutions: ^3H, ^{14}C, and ^{32}P.

Response to Different Radiations

The relative scintillation yield is defined as the photon yield (or pulse height response) per unit energy deposited in the scintillator solution. This yield depends not only on the energy of the particle but on the type of particle causing the production of excited molecules. The yield for electrons is almost directly proportional to the energy of the electron. Figure V-1

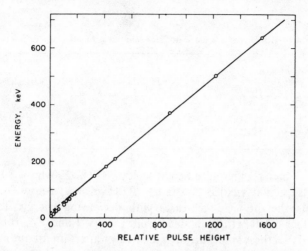

Fig. V-1. Pulse height–energy relationship for electrons, 6.5–636 keV.

shows the relative pulse height response as a function of the energy of the electron for relatively high-energy electrons (> 80 keV). The relationship is described by the equation

$$\text{relative pulse height (RPH)} = k(\text{Energy (MeV)} - 0.018)$$

where k is a normalization factor which depends on the gain of the overall system and 0.018 is the intercept if the straight line part of the plot is extrapolated to zero RPH value (1–3).

The pulse height response is not linear below about 80 keV electron energy (3), as shown in Fig. V-2. The measure of the relative response can

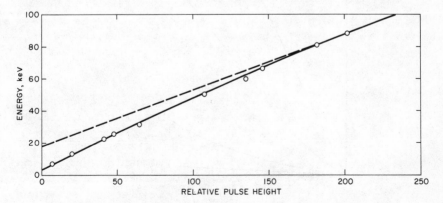

Fig. V-2. Pulse height–energy relationship for low-energy electrons.

Table V-1

Specific response (dRPH/dE) of a liquid scintillator solution as function of electron energy

Electron energy (keV)	dRPH/dE	Normalized
> 80	2.8	1.00
50	2.2	0.79
20	2.0	0.71

be related to the specific pulse height yield, dRPH/dE. Above 80 keV the specific pulse height yield is constant: dRPH/$dE = k$; below 80 keV the specific pulse height yield decreases with decreasing energy. Table V-1 lists the values of dRPH/dE obtained from Figs. V-1 and V-2. It should be remembered that these values are only relatively important for one measuring system, depending on the gain, optical system, and other factors which will affect the slope of the energy-versus-pulse height plot.

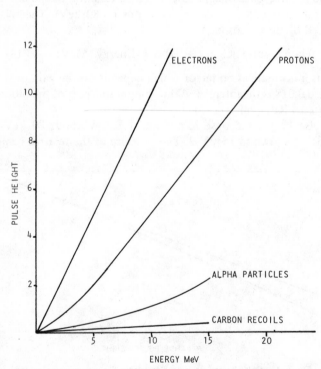

Fig. V-3.   Pulse height relationships for electrons, protons, alpha particles, and recoil carbon atoms (4).

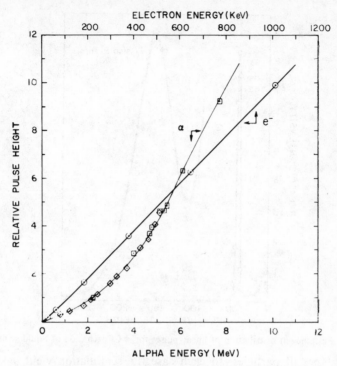

Fig. V-4. Pulse height–energy relationship for alpha particle and electron excitation of a liquid scintillator (1).

Figures V-3 and V-4 show the response of a scintillator solution to different energies of alpha particles and protons. The higher specific ionization of the alpha particles and protons is reflected in the lower scintillation yield per unit energy. More of the energy goes into nonlight-producing events.

In one case the energy released by fission was measured by a liquid scintillator solution (5). The total energy released by the spontaneous fission of ^{252}Cf was absorbed in a liquid scintillator solution and the pulse height response was measured. Since each fission releases 180 MeV of energy, the fission process will produce a peaked distribution of pulse heights as shown in Fig. V-5. A point of reference is produced by the 6.1-MeV alpha particles produced by the radioactive decay of ^{252}Cf.

The relative scintillation yield for different particles varies considerably with the type of particle and its specific ionization. The higher specific ionization gives the lower scintillation yield. The 180-MeV fission events produce the same number of photons as an electron with 1.3% of that energy, 2.34 MeV. Table V-2 compares the relative scintillation yields for

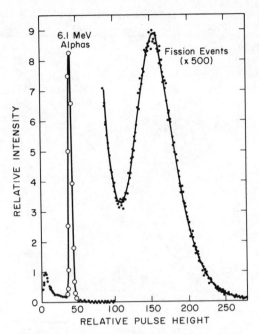

Fig. V-5. Pulse height distribution of linear pulses for ^{252}Cf excitation of liquid scintillator.

different types of particles. In each case the scintillation yield has been normalized relative to an electron.

Counting Beta Emitters

Beta emitters include both negatron and positron emitters. Both types of particles, negatively charged electrons and positively charged electrons, interact with the liquid scintillator in the same manner (6, 7). They produce a continuum of pulse heights corresponding to the continuum of beta energies, zero energy to E_{max}. The only difference is that when the positron is completely stopped, i.e., at zero kinetic energy, two γ rays of 0.51-MeV energy are produced. For most small-volume liquid scintillator solutions (20 ml), the two γ rays are not detected. Figures V-6 and V-7 show typical pulse spectra obtained for a negatron and a positron emitter.

Counting Other Electron Emitters

Two other types of radioactive decay also produce electrons, internal conversion and electron capture. In both of these the electrons produced

Table V-2

Relative scintillation yield per million electron volts for different types of particles exciting a liquid scintillator

Particle	Relative scintillation yield[a]
Electron	1.00
Proton	0.50
Alpha[b]	0.12–0.08
Fission[c]	0.013

[a] Electron with this fraction of particle energy will give same scintillation yield.
[b] For alpha-particle energies from 7.0 (0.12) to 4.0 MeV (0.08).
[c] For ^{252}Cf spontaneous fission, 180 MeV.

are monoenergetic or groups of monoenergetic electrons as opposed to the energy continuum of electrons produced by beta decay. Thus the pulse distribution will be a peak distribution as shown in Fig. V-8 for the conversion electrons from the decay of ^{113m}In.

Internal Conversion

The internal conversion process is an alternative and competitive process to the γ-ray mode of decay. The energy which normally appears in the form of the γ ray is imparted to an extranuclear electron. The γ ray is *not* emitted and then absorbed by the electrons. The electrons most likely to be emitted are those closest to the nucleus which have a binding energy less than the energy of the gamma transition.

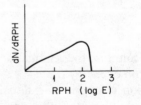

Fig. V-6. Pulse distribution for a negatron (β^-) emitter, ^{45}Ca, dissolved in a liquid scintillator solution ($E_{max} = 255$ keV).

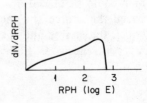

Fig. V-7. Pulse distribution for a positron (β^+) emitter, ^{22}Na, dissolved in a liquid scintillator solution ($E_{max} = 540$ keV).

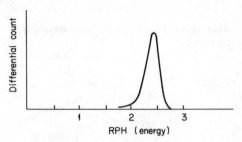

Fig. V-8.   Pulse distribution for 369-keV conversion electrons of ^{113m}In.

The electron will have an energy equal to the difference between the energy of the gamma transition, E_γ, and the binding energy of the electron BE_e:

$$E_e = E_\gamma - BE_e.$$

After the electron has been ejected, the atom is left with a vacancy in the shell where the conversion occurred. This vacancy is then filled by a series of extranuclear electron rearrangements which produce X rays and Auger electrons. All of these extranuclear processes occur in a very short time ($< 10^{-11}$ sec) and therefore appear as coincident with the conversion electron. Thus the total energy that the liquid scintillator solution will absorb will be the sum of all the processes that are stopped, or are degraded, within the solution. The difference between the E_γ and the energy deposited in the scintillator solution is equal to any X rays which escape from the scintillator solution without interacting with the solution.

After all of the atomic rearrangements following the filling of the vacancy left by the conversion electron, the total energy of the conversion electron, the X rays, and Auger electrons will equal the E_γ:

$$E_\gamma = E_e + E_{X\,rays} + E_{Auger}.$$

Thus if all of these are stopped in the scintillator solution, the scintillation yield would be the same as stopping a single electron of energy E_γ.

Early experimenters used high-energy conversion electrons as external sources for the evaluation of the scintillation yield of different solutions and as a method of quench measurement (8–10). A source of ^{137}Cs–^{137m}Ba was placed on the cap of a sample bottle or a disk which could be placed over the scintillator solution with nothing but air between the source and the solution. Even the smallest mass of material between the source and the solution would absorb or greatly degrade the conversion electrons. Some typical spectra obtained by this method are shown in Fig. V-9.

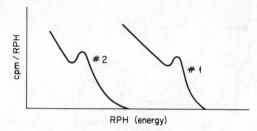

Fig. V-9. Differential pulse spectra for two different solutions excited by the conversion electrons from an external ^{137}Cs–^{137m}Ba source.

The peak occurs at the highest RPH value for the solution with the greatest scintillation yield or the minimum amount of quench (corresponding to spectrum 1 of Fig. V-9). If another solution has a lower scintillation yield or a higher degree of quench, the peak will occur at a lower RPH value (corresponding to spectrum 2 of Fig. V-9).

Since the source of electrons is outside the scintillation solution, surface effects cause the peaks to be much broader than for a dissolved sample. The conversion electrons penetrate only a short distance below the surface of the solution before coming to rest. The light produced is isotropically emitted. The photons have to be transferred completely through the solution before being collected on the face of the multiplier phototube. Photon absorption by the components of the solution, even though very small, will cause a greater variation in the number of photons that reach the multiplier phototube per unit energy.

Another use of undissolved sources of conversion electrons was that of a ^{57}Co source on a needle (11, 12). The needle was inserted into the scintillator solution for measure of the scintillation efficiency. There are two transitions in the radioactive decay of ^{57}Co which give rise to conversion electrons, those occurring at 14 and 122 keV. The quench and scintillation yield were measured by the change in observed count rates of the ^{57}Co source in two counting channels; one channel was set to count conversion electrons which produced pulses between zero and the pulse equivalent to 14-keV electrons, and the second channel to count electrons which produced pulses above the pulse equivalent to 14 keV and up to the pulse equivalent of 122-keV electrons. Figure V-10 shows the general pulse distribution and channel settings.

When the radionuclide is dissolved in the scintillator solution to give a homogeneous solution, the detector system has essentially a 4π geometry. The total energy for excitation of the scintillator solution is the sum from all the processes by which electrons and/or X rays are stopped or slowed

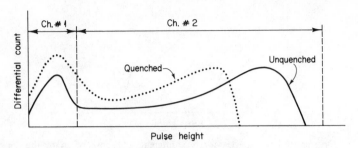

Fig. V-10. Pictorial scheme of quench monitor with [57]Co-plated needle inserted into scintillation solution (11, 12).

down in the solution. For nuclides that decay by internal conversion the total includes energies from the conversion electrons, Auger electrons, and X rays. The energy not available for excitation is that carried off by those X rays which escape from the solution and electrons which strike the container walls while still possessing kinetic energy. For scintillator solutions of 15 to 20 ml only K X rays of low atomic energy nuclides are energetic enough to escape. For high atomic number nuclides the escape of both K and L X rays may be energetically possible. Figure V-11 shows the pulse spectrum obtained for [131m]Xe which decays predominantly by the conversion electron emission.

Electron Capture

Radioactive decay by electron capture leads to the production of Auger electrons. The yield of Auger electrons is equal to one minus the fluorescence

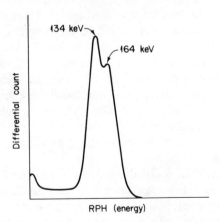

Fig. V-11. Pulse distribution for [131m]Xe conversion electrons when Xe is dissolved in the scintillator solution.

yield, $\omega_{Auger} = 1 - \omega_X$. For a vacancy in the K shell of a nuclide, the fluorescence yield ω_K is the probability that a K X ray will be formed. When the K X ray is not formed the energy is transferred to an electron, and the electron will have an energy equal to the difference between the energy of the X ray and the binding energy of the ejected electron:

$$E_{Auger} = E_{K\,X\,ray} - E_{binding}.$$

The ejection of an Auger electron will in turn leave a vacancy in one of the orbital electron shells. The subsequent rearrangement will lead to the production of more Auger electrons and characteristic X rays. In those cases where the secondary X rays are low energy and thus totally absorbed in the scintillator solution, the total energy deposited in the scintillator solution will be the binding energy of the initial captured electron.

Consider the radionuclide ^{55}Fe which decays 100% by electron capture. Ninety percent of the captures occur from the K shell and 10% from the L shell (13). Upon filling the K shell, 6.5 keV of energy is released in the form of a K X ray (5.9 keV) and/or Auger electrons; upon filling the L shell, 0.77 keV of energy is released in the form of an L X ray (0.69 keV) and/or Auger electrons. The very low-energy X rays will be stopped in very small volumes of solution almost quantitatively. Figure V-12 shows the pulse spectrum of a ^{55}Fe sample (organic phosphoric acid complex) in a liquid scintillator solution of 0.25 ml total volume and measured on a single multiplier phototube counter (3, 14). The spectrum shows two peaks, one at 6.5 keV corresponding to the K electron binding energy, and one at lower energy corresponding to the energy equivalent of a single photoelec-

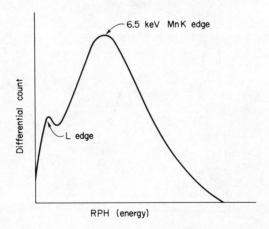

Fig. V-12. Pulse spectrum for ^{55}Fe source dissolved in liquid scintillator.

tron. The L edge energy is low enough so that it has only a finite probability to produce at most a single photoelectron.

Another radionuclide which decays 100% by electron decay is ^{125}I. The spectrum obtained with an aqueous Na ^{125}I solution in an emulsifier counting system is shown in Fig. V-13. The two peaks correspond to two energy bands of approximately 12 and 40 keV. The ^{125}I does not electron capture directly to the ground level of ^{125}Te. The 35-keV excited level of ^{125}Te decays predominantly by internal conversion.

The radionuclide ^{51}Cr also decays 100% by electron capture. In 9% of the electron captures an excited energy level of ^{51}Ti is formed. The 320-keV level decays by the emission of a γ ray. Thus in the pulse height spectrum shown in Fig. V-14 the long plateaus at the higher pulse heights are due to Compton scattered electrons produced in the scintillator solution by the 320-keV γ ray. The lower energy band is due to the 5-keV energy released upon filling the Ti K shell vacancy.

Counting Alpha Emitters

It was first demonstrated in 1954 that it is feasible to count alpha particles in a liquid scintillator solution (15). Alpha-emitting nuclides emit groups of monoenergetic alpha particles. There are nuclides that emit only one energy alpha particle, while there are others that emit as many as 20 different monoenergetic groups of alpha particles. Most of the naturally occurring alpha-emitting radionuclides emit at least two monoenergetic groups. In most

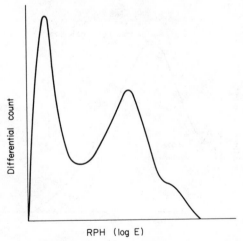

Fig. V-13.   Pulse spectrum for ^{125}I.

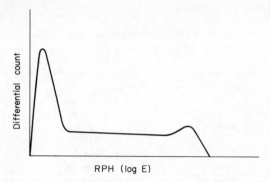

Fig. V-14. Pulse spectrum for ⁵¹Cr.

cases there are two groups which comprise nearly 100% of the alpha particles; the rest of the groups usually make up less than 1% of the total alpha particles. A general decay scheme is shown in Fig. V-15. About 70% of the decays lead directly to the ground state of the daughter by emission of an alpha particle of energy E_{α_1}. The remaining 30% decay by emission of an alpha particle that is 30–50 keV (depending on the individual radionuclide) less energetic than E_{α_1}. Table V-3 lists several radionuclides, and the energies and relative abundances of the most prominent energy groups, as well as the differences between them. The transition from the excited level to the ground level occurs almost 100% by internal conversion with the emission of a monoenergetic electron which is coincident with the alpha particle that leads to the excited level.

The distribution of pulses from a sample of ²³⁸Pu dissolved in a liquid scintillator solution (2) is shown in Fig. V-16. Two pulse distributions are given, the differential and the integral. The differential distribution was obtained by measuring the counts that occur in a narrow pulse height band as this band is varied over the total pulse height spread. The integral distribution is the total number of pulses that exceed a given pulse height

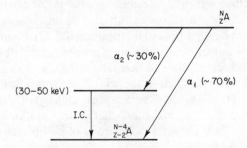

Fig. V-15. Generalized decay scheme for an alpha-emitting nuclide.

Table V-3

Alpha-emitting nuclides and the relative abundances of their major alphas

Nuclide	Alpha energy (MeV)	Relative abundance of the alphas[a] (%)	ΔE (MeV)
^{238}Pu	5.495, 5.452	72, 28	0.043
^{236}Pu	5.763, 5.716	69, 31	0.047
^{238}U	4.195, 4.147	77, 23	0.048
^{234}U	4.768, 4.717	72, 28	0.051
^{232}Th	4.007, 3.948	76, 24	0.059
^{230}Th	4.682, 4.615	76, 24	0.067
^{226}Ra	4.777, 4.589	94, 6	0.188
^{221}Fr	6.33, 6.12	84, 16	0.21
^{239}Pu	5.147, 5.134, 5.096	72.5, 16.8, 10.7	0.013, 0.038
^{210}Po	5.305	100	0
^{147}Sm	2.18	100	0

[a] Only the more abundant alphas are listed. Those with abundances less than 1% are omitted.

value and is measured at different pulse height values over the total spread. The alpha particles are counted with 100% efficiencies as long as the sample is dissolved in the scintillator solution. If the sample were plated on the walls or suspended on a solid matrix, the efficiency would be less than 100%. Also the differential distribution would be much wider and would show pulses below the peak at all values to zero pulse height, i.e., there is a tail on the distribution as shown in Fig. V-17.

Two or more different alpha emitters can be measured simultaneously in a liquid scintillator solution (2, 16–19). Each can be determined independently if the energy difference between the two groups is sufficient to be resolved. Figure V-18 shows the pulse distribution of a sample containing both ^{233}U and ^{238}Pu. The main alpha energy groups of ^{233}U and ^{238}Pu are 4.8 and 5.5 MeV, respectively. Each can be stripped independently of the other, and the disintegration rate of each radionuclide can be determined.

The alpha-particle-produced pulses can also be distinguished from those of a beta continuum by their peaked distribution (20, 21). Figure V-19 shows the distribution of pulses obtained from a sample of ^{210}Pb–^{210}Bi–^{210}Po dissolved in a liquid scintillator solution. The sample was an aqueous dilute HCl solution and was emulsified in the scintillator solution with the aid of an emulsifier like Triton X-100. The 5.3-MeV alpha particles from ^{210}Po can be distinguished easily from the low-energy beta continuum ($E_{max} = 14$ and 61 keV) and 47-keV conversion electrons of ^{210}Pb and the beta continuum ($E_{max} = 1.15$ MeV) of ^{210}Bi.

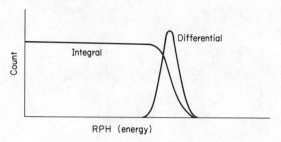

Fig. V-16. Differential and integral pulse spectra for ²³⁸Pu dissolved in a liquid scintillator.

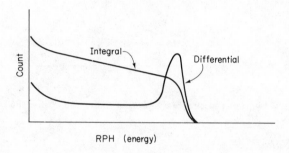

Fig. V-17. Effect of part of sample being on solid matrix upon the differential and integral spectra of ²³⁸Pu.

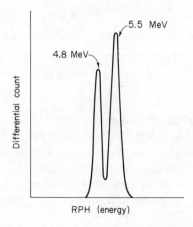

Fig. V-18. Pulse spectra of sample containing both ²³³U and ²³⁸Pu with a single MPT counter (2).

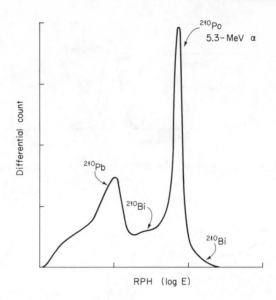

Fig. V-19.   Pulse spectra for sample containing ^{210}Pb–^{210}Bi–^{210}Po.

Alpha–Beta Ratio

The alpha–beta ratio (α/β) has been defined in two different ways. In one case it is defined as the ratio of the energies of the alpha particle and the electron which produce the same pulse height (PH) response:

$$\alpha/\beta = E_\alpha/E_e \tag{1}$$

at the same PH. The other definition relates the ratios of pulse heights divided by the energies of the particles:

$$\frac{\alpha}{\beta} = \frac{PH_\alpha/E_\alpha}{PH_e/E_e} \tag{2}$$

In each case E_α and E_e are the energies of the alpha particle and electron, respectively, and PH_α and PH_e are the pulse height responses from the alpha particle and electron, respectively. Since these two definitions give different ratios, it is important to state the method of obtaining the α/β ratio. The difference in these two ratios is due to the nonproportional relationship between alpha-particle energy and pulse height response (see Fig. V-3).

The α/β ratio has been used to study various scintillator solutions for evaluation of the solvents and solutes (22). The scintillation response for the two types of particles can be used to study effects on decay times and on the formation of free radicals (G_r value). The α/β ratio decreases with increasing solute decay time for several different solutes in benzene; it also decreases for solvents that have a greater susceptibility for the formation of free radicals.

Using the definition given in Eq. (1) the α/β ratio has been obtained for alpha-particle energies ranging from 0.79 to 7.68 MeV (1). Table V-4 lists the energy of the alpha particle and electron that produce the same pulse height response and the α/β ratio. The ratios vary from ~ 23 for the lowest energy alpha particles to ~ 8 for the highest energy alpha particles. It is often quoted that an alpha particle produces a pulse height response equivalent to an electron with 0.10 of the alpha-particle energy. This is only true for an alpha particle of ~ 5.3 MeV energy.

The α/β ratio is also subject to changes due to changes in solute concentrations and temperature variations (23). The actual relationship between the α/β ratio and temperature or solute concentration is not known but appears to have a complicated interpretation.

Table V-4

Equal response values of alpha and electron energies and values of α/β ratios

Alpha energy (MeV)	Electron energy (MeV)	α/β
1	0.044	22.7
2	0.100	20.0
3	0.176	17.0
4	0.295	13.6
5	0.450	11.1
5.3	0.530	10.0
5.8	0.590	9.9
6	0.630	9.5
7	0.840	8.4
8	1.000	8.0

Energy Resolution

The resolution of a counting system is a measure of the ease of obtaining different and distinguishable responses for two like particles with small differences in their energy. The resolution is usually expressed in one of two forms, the line width or the energy resolution. The line width is obtained from the equation:

$$\text{line width} = (\text{FWHM})/\text{PH}_{\max} \qquad (3)$$

where FWHM is the full width at one-half the maximum of the peaked distribution in pulse height units (discriminator divisions, channels, etc.). Figure V-20 illustrates the method for obtaining the FWHM value.

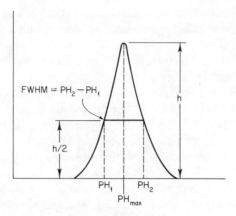

Fig. V-20.   Method of obtaining line-width value. Line width = FWHM/PH$_{max}$.

The energy resolution is obtained by converting the FWHM value into the equivalent energy of the pulse height difference. The energy resolution is given by the relation:

$$\text{energy resolution} = \text{FWHM (in MeV)}/E \text{ (in MeV)}. \qquad (4)$$

For alpha particles the line width and energy resolution will have different values due to the nonproportional relationship between the pulse height response and the alpha-particle energy. The energy resolution will be essentially constant (at least for the energy range 4–7 MeV) because the energy–pulse height relationship is linear (2) (see Fig. V-3). Table V-5 lists the alpha-emitting isotope, alpha-particle energy, energy resolution, and line width obtained with a single multiplier phototube counter.

Table V-5

Line width and energy resolution as a function of alpha energy

Isotope	Alpha energy (MeV)	Energy resolution	Line width
^{232}Th	4.0	0.058	0.129
^{236}U	4.5	0.057	0.120
^{233}U	4.8	0.058	0.114
^{239}Pu	5.1	0.059	0.112
^{238}Pu	5.5	0.058	0.105
^{236}Pu	5.75	0.058	0.102
^{242}Cm	6.1	0.058	0.099
^{220}Rn	6.3	0.057	0.095
^{216}Po	6.8	0.058	0.089
^{217}At	7.0	0.059	0.084

In early reported work commercial instruments have been shown to give much poorer resolution of alpha particles than those listed in Table V-5. Most investigators have reported line widths of 24–50% for approximately 5-MeV alpha particles (19, 24–27). Recent studies indicate that most of the poorer results were due to the type of sample being measured (19, 28). Most early reports utilized aqueous solutions of the alpha emitter. Figures V-21 and V-22 show the resolution obtained with a commercial coincidence-type instrument. In this case the samples were prepared as complexes of an organic phosphoric acid and the complex dissolved directly in the toluene which served as the solvent of the scintillator solution. Under these conditions there was essentially no quenching.

The pulse height–energy relationship obtained in the commercial instrument (Beckman LS-250) is shown in Fig. V-23 and is essentially identical to the corresponding part of Fig. V-3 when converted to a log energy scale. Conversion of the FWHM to an energy equivalent showed an energy resolution of about 10% for 5.5-MeV alpha particles.

Figure V-24 shows a comparison of the line width obtained with ^{241}Am alpha particles in different scintillator solutions. Curve (a) was obtained with an ^{241}Am–organic phosphoric acid complex (i.e., a homogeneous solution) in a toluene base liquid scintillator solution. Curve (b) was ob-

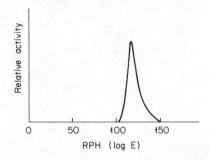

Fig. V-21. Pulse spectrum for ^{241}Am (5.5 MeV).

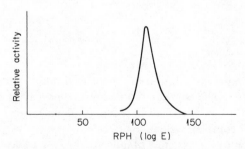

Fig. V-22. Pulse spectrum for normal uranium: ^{238}U (4.2 MeV) and ^{234}U (4.7 MeV).

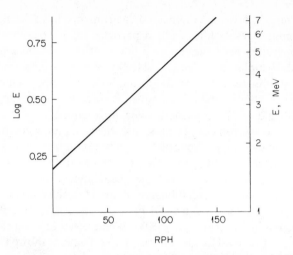

Fig. V-23. Energy–pulse height relationship obtained with a commercial instrument with logarithmic amplification.

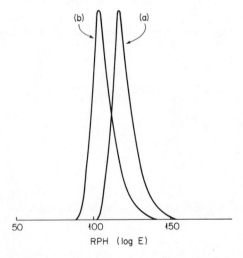

Fig. V-24. Pulse spectra for [241]Am samples in (a) homogeneous toluene solution, (b) aqueous–emulsifier–toluene solution.

tained with an aqueous solution of [241]Am in a liquid scintillator solution with an emulsifier. The emulsifier–aqueous system gave a greater FWHM value and a lower light yield per unit energy of excitation than the organic phosphoric acid complex system.

The resolution of the system will indicate the ease of counting two alpha emitters in the same solution and still being able to separate the pulse due

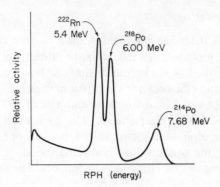

Fig. V-25. Pulse spectrum for ^{222}Rn and daughters dissolved in toluene scintillator solution with a single MPT counter.

to each alpha emitter. Figure V-25 shows the pulse height distribution obtained for a sample of radon gas dissolved in a liquid scintillator solution with a single MPT counter (17). The ^{222}Rn decays by emission of 5.49-MeV alpha particles. Several radioactive daughters are produced before the decay chain finally reaches the long-lived ^{210}Pb ($t_{1/2} = 22$ years). The decay chain is shown in Fig. V-26. Three alpha groups, of energies 5.49, 6.00, and 7.68 MeV, and three beta continuums are produced, all of which can be observed from examination of the pulse height distribution.

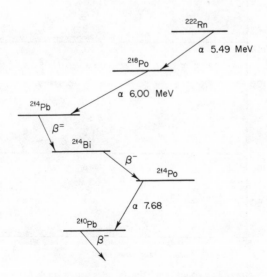

Fig. V-26. Decay scheme of ^{222}Rn and daughters.

Low-Level Alpha Counting

Because alpha particles are monoenergetic a narrow counting channel can be selected to include only the alpha-produced pulses. The biggest advantage to this is the low backgrounds that can be obtained while maintaining 100% counting efficiency for the alpha particles. Figure V-27 shows the pulse distributions obtained for three samples of ^{238}Pu, which decays by emission of 5.5-MeV alpha particles. The background was a smooth continuum and the alpha-particle-produced pulses appeared as a peak. The background in the window that bracketed the alpha peak was 0.26–0.30 cpm (18). Samples with as little as 0.22 dpm could be determined easily. This corresponds to only 5.7×10^{-15} g of ^{238}Pu.

The moderate energy resolution of the liquid scintillator in a single MPT system allows one to determine the disintegration rate of two or more different alpha emitters in the same sample, provided their energy difference is great enough to allow complete separation of the pulse produced by each alpha emitter. Figure V-28 shows the results obtained using

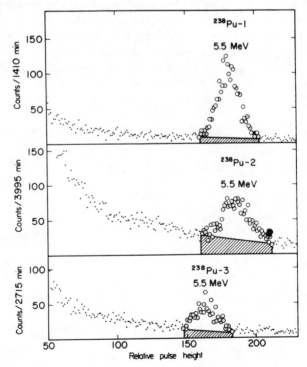

Fig. V-27. Pulse spectra of low count rate ^{238}Pu samples with a single MPT counter showing the low background and sensitivity.

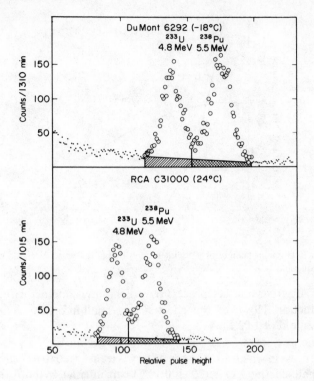

Fig. V-28. Pulse spectra for single sample containing both ^{238}Pu and ^{233}U showing simultaneous determination of count rate of each.

two different counting systems; one system utilized a single DuMont 6292 multiplier phototube at $-18°C$ and the other system utilized a single RCA C31000 (bialkali) multiplier phototube at room temperature. The sample contained 1.6 dpm of ^{233}U (4.8-MeV alpha particles) and 2.2 dpm of ^{238}Pu (5.5-MeV alpha particles). In both cases the activity of the two alpha emitters was determined with excellent agreement (18).

Counting Gamma Emitters

Gamma rays can be counted in a liquid scintillator because of their interactions with the materials that make up the solution. These interactions produce electrons which excite the scintillator solution like a beta particle, conversion electron, etc. The γ-ray interactions occur mainly by Compton scattering or photoelectric effect. The probability of Compton scattering and photoelectric effect varies with the energy of the γ-ray as shown in

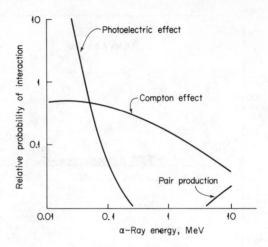

Fig. V-29. Relative probability of γ-ray interaction by photoelectric effect, Compton effect, or pair production.

Fig. V-29. There is another process by which γ rays interact with matter—pair production. However, this effect is very small in organic material and has a threshold of 1.02 MeV.

Low-energy γ rays (<20 keV) interact predominantly by the photoelectric process in which all the γ-ray energy is transferred to a single electron. Intermediate-energy γ rays (20–100 keV) can interact by both the photoelectric and Compton processes. In the Compton process only part of the γ-ray energy is imparted to an electron and the remaining energy is in the form of a new γ ray. Higher energy γ rays (0.1–3 MeV) interact almost exclusively by the Compton process. There is no fine demarcation between these processes. Many γ rays will undergo several steps of interactions. A moderate-energy γ ray may undergo only a single Compton scattering before leaving the scintillator solution. It may also undergo multiple Compton scattering; i.e., the new γ ray may have a Compton scattering, and it may also undergo a photoelectric interaction. The response of the scintillator solution will be equal to the amount of energy from the initial γ ray that is converted to electron energy in the material of the solution.

The probability of a γ-ray interaction with any matter is a function of, among other things, the electron density of the matter. Since the liquid scintillator solution is made up of atoms of low atomic number [H (1), C (6), N (7), and O (8)], the probability of a γ-ray interaction is small, which means that the counting efficiency for γ rays in small volumes of liquid scintillators is low. The greater the volume of the scintillator solution, the greater is the probability of the γ-ray interaction. Thus very large volume

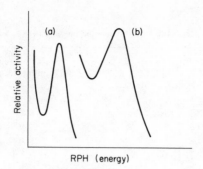

Fig. V-30. Pulse spectra produced in a liquid scintillator solution by external γ-ray sources. (a) 22.1-keV AgK X rays; (b) 59 keV γ rays from ^{241}Am.

liquid scintillation counters, i.e., a whole body counter, are quite efficient for γ rays.

External sources of γ rays have been used in liquid scintillation counting for many years as a monitor of the quench level. In 1958 (25) the 59-keV γ rays of ^{241}Am and the 22-keV K X rays of Ag from a ^{109}Cd$-^{109m}$Ag source were used to check the level of quench of scintillator solutions. These low-energy γ rays were measured by the photoelectric effect by which the energy of the γ rays or K X rays was totally deposited in the liquid scintillator solution. Figure V-30 shows the pulse height spectra obtained with a scintillator solution excited with these low-energy γ-ray sources. The peaked parts of the spectra correspond to those interactions which result in the total energy of the γ ray being deposited in the solution.

Higher-energy γ rays are detected in liquid scintillator solutions by the Compton scattered electrons they produce (29). Because of the relatively small volume (≤ 20 ml) of solution the probability of more than a single scattering of the γ ray is very small. The energy distribution of the electrons produced by single Compton scattering is predicted by the simple equations of conservation of momentum and kinetic energy. The maximum energy that a scattered electron can have occurs when the γ ray is scattered at 180° from the direction of the initial γ ray. The equation for the maximum electron energy, E_{max}, is dependent only on the energy of the initial γ ray:

$$E_{max} = 2E_\gamma^2/(2E_\gamma + 0.51)$$

where E_γ is the energy of the γ ray in million electron volts and 0.51 is the rest mass of the electron. Some E_{max} values of various γ-ray energies are listed in Table V-6.

The Compton scattered electrons can have any energy between zero and E_{max}. Theoretically the probability is equal for any energy electron being

Table V-6

Maximum energy of Compton scattered electrons for single scatter as a function of γ-ray energy

γ-Ray energy (MeV)	E_{max} (MeV) of Compton electrons
0.1	0.028
0.2	0.088
0.3	0.162
0.4	0.244
0.5	0.331
0.6	0.421
0.8	0.607
1.0	0.797

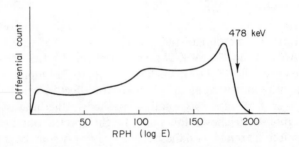

Fig. V-31. Pulse spectrum produced in logarithmic amplification system by Compton scattered electrons produced by 662-keV γ rays from ^{137}Cs–^{137m}Ba.

produced. The probability of slightly different response for equal energy deposited in the liquid scintillator and the response threshold of the detector and other factors leads to the logarithmic response spectrum of Compton electrons as shown in Fig. V-31. The same spectrum obtained with a linear energy response system is shown in Fig. V-32. Figure V-33 shows the pulse spectrum obtained for ^{22}Na which decays by producing both 0.51- and 1.28-MeV γ rays. Each γ ray produces its own Compton scattering distribution. The E_{max} of each Compton distribution can be easily observed. The energy associated with the given Compton edge, E_{max}, has been used to calibrate the counting system for pulse height–energy relationships. In most cases the selection of the pulse height corresponding to E_{max} is very important. If the pulse height corresponding to the half-height of the Compton edge, PH_1 of Fig. V-34, is selected as equal to E_{max}, the response is about 4% higher than that obtained for electron sources dissolved in the liquid scintillation solution (1). The difference is probably due to the method of selecting the pulse height corresponding to E_{max}. If E_{max} were

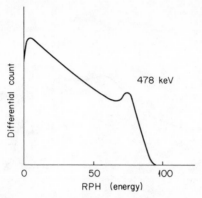

Fig. V-32. Pulse spectrum produced in linear amplification system by Compton scattered electrons produced by 662-keV γ rays from ^{137}Cs–^{137m}Ba.

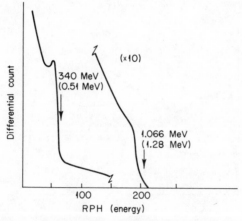

Fig. V-33. Pulse spectrum (linear amplification) produced by Compton scattered electrons produced by the 0.51- and 1.28-MeV γ rays from ^{22}Na.

selected as equal to the small peak prior to the sharp edge, PH_2 of Fig. V-34, the agreement would be much better. The half-height method is used more often because of its sensitivity. The edge is changing so rapidly that the pulse height corresponding to $h/2$ can be obtained within very small limits, whereas the peak is somewhat more difficult to locate. Also, as the amount of quench increases the edge remains fairly sharp but the peak becomes broader and the pulse height corresponding to the peak is known less accurately, as shown in Fig. V-35.

Gamma-ray counting efficiency in small volumes (≤ 20 ml) of liquid scintillator solution is usually low. Table V-7 lists counting efficiencies for a few γ-ray emitters with 20 ml of a standard toluene base liquid scintillator

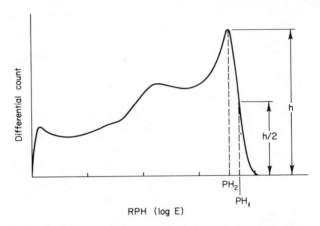

Fig. V-34.   Method of obtaining pulse height calibration of the Compton edge. PH_1 corresponds to one-half the maximum count.

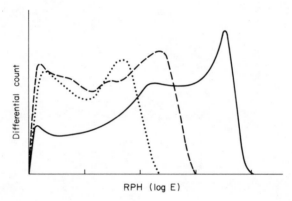

Fig. V-35.   Effect of quenching on the pulse distribution for Compton electrons produced by 662-keV γ rays from ^{137}Cs–^{137m}Ba. Increasing quench shifts spectrum to lower relative pulse height.

solution. The lower energy γ rays are counted with the highest efficiency because there is a greater probability that they will be scattered in the scintillator solution than will the higher energy γ rays.

The probability of a γ ray being scattered or stopped by a given material is dependent on the electron density of the material. The low atomic numbers of the atoms that make up the scintillator solutions [H (1), C (6), N (7), O (8)] provide a material with a very low electron density. If a material could be added to the solution to increase the electron density without drastically reducing the scintillation yield (i.e., without a large amount of quenching), the gamma counting efficiency could be improved. In Table

Table V-7

Counting efficiencies of γ and X rays

Nuclide	Energy of most prominent γ-rays (MeV)	Typical efficiency (%)	
		In standard solution	In Pb-loaded solution
[125]I	0.027, 0.055[a]	10	50
[131]I	0.365, 0.640	7	10
[60]Co	1.17, 1.33	3	3

[a] Sum of two 0.0275-MeV γ rays.

V-7, the last column lists the counting efficiency when tetrabutyllead was added to the scintillator solution to give a concentration of 5% tetrabutyllead by volume. The greatest increase in counting efficiency is with the low-energy γ rays; the high-energy γ rays still have low counting efficiencies. The 5% tetrabutyllead decreased the scintillation yield by only a few percent. Figure V-36 shows the effect of various amounts of tetrabutyltin upon the counting efficiency (30).

Several different vials have been designed to make it possible to count gamma emitters in commercial liquid scintillation counters. Figure V-37

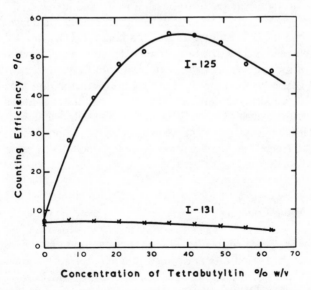

Fig. V-36. Relationship between concentration of tetrabutyltin and gamma counting efficiency of [125]I (○) and [131]I (×), for 15 ml toluene solution containing 5 g/liter PPO and 0.1 g/liter POPOP (30).

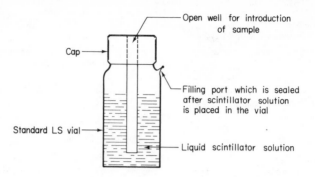

Fig. V-37.   Bottle for counting gamma emitter with loaded scintillator solutions.

shows one type of such vial (30). The sample of gamma emitter is placed in a small container which is slipped into the open well. The vial is placed in the sample changer of a standard liquid scintillation counter and automatically descends into the counting chamber for counting. Caution should be observed that the gamma sample vial does not extend unduly above the top of the standard liquid scintillator (LS) vial. If it does, it is possible that the light shutter on automatic sample changers will catch the small vial, causing possible breakage with subsequent contamination, inaccurate counting, and light leakage into the phototube chamber.

Different types of compounds have been used as electron density increases in liquid scintillator solutions for gamma counting. Organic soluble compounds have been used with aromatic solvent systems (30, 31). Other aqueous solutions of inorganic salts have been used with the newer emulsifier solutions. In all cases it should be checked to confirm whether the added material adversely affects the scintillation yield of the scintillator solution. A determination of the external standard channels ratio will give a measure of the relative scintillation yield. A simple external standard gross count rate will not be valid since the counts will be increased due to the greater stopping power of the solution.

Neutron–Proton Counting

Neutrons are detected in a liquid scintillator solution in one of two ways, scattering by hydrogen atoms or absorption by certain isotopes. The fast neutrons, which have energies greater than a few kiloelectron volts, are elastically scattered by the protons (H atoms) in the material. The scattering process is like the billiard ball principle since the mass of the neutron and that of the proton are essentially equal. The recoil proton can have any

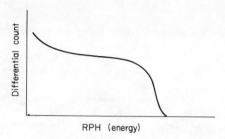

Fig. V-38. Pulse spectrum of recoil protons produced by scattering of monoenergetic neutrons in a liquid scintillator solution.

energy from zero to the total energy of the neutron, depending on the angle of the collision. The protons are ionizing particles which produce excited molecules in the scintillator solution which in turn lead to scintillations. The protons will produce a scintillation yield equal to an electron with one-half of the proton energy. Thus a monoenergetic neutron will produce a continuum of protons, and thus a continuum of pulse intensities from the scintillation detector. Figure V-38 shows a typical spectrum obtained with a source of monoenergetic neutrons (32). If the energy of two neutron groups is sufficiently different, it is possible to obtain a spectrum as shown in Fig. V-39. The relative number of each group is proportional to the height of the edge (33).

The neutron counting efficiency of the scintillator solution is a function of a combination of the volume of the scintillator solution and the energy of the neutrons. There is a threshold energy which will be a factor in the neutron counting efficiency. The threshold will vary with the type of scintil-

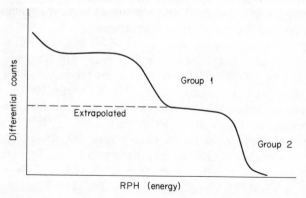

Fig. V-39. Pulse spectrum of recoil protons produced by scattering of two different energy groups of monoenergetic neutrons in a liquid scintillator solution.

Table V-8

Neutron counting efficiency as a function of neutron energy and scintillator volume[a]

Neutron energy (MeV)	Efficiency (%) for cylinder of scintillator with dimensions	
	2-in. diameter 2.5 in. long	3-in. diameter 4 in. long
0.7	73.6	
1.0	68.2	80.5
2.0	55.5	
4.0	46.6	58.8
8.0	31.7	
14.0	24.6	34.1

[a] From R. Batcher, W. B. Gilboy, J. B. Parker, and J. H. Towle, *Nucl. Instrum. Methods* **13**, 70 (1961).

lator solution and the quench level. Table V-8 lists counting efficiency of monoenergetic neutrons for two different sizes of scintillator solutions.

High-energy neutrons have been counted with high efficiency by the n, γ reactions in gadolinium-loaded, mineral oil scintillator solutions (34). The Gd is complexed as an organic acid and dissolved in a mixture of mineral oil and tri-*n*-octylphosphine oxide with scintillator solutes. A counter with 1 m³ volume loaded with 5 g/liter of Gd gave a counting efficiency of 80% for 10-MeV neutrons.

Thermal neutrons, which have energies much less than an electron volt, do not possess enough energy to be detected by the elastic scattering of protons. To measure thermal neutrons the liquid scintillator solution is "loaded" with an isotope that has a very high probability of interaction with the neutrons (cross section) with a resulting nuclear emanation which is efficiently counted by the scintillator solution.

Scintillator solutions loaded with boron which has been enriched in ^{10}B have been used with high efficiencies for measuring thermal and near-thermal neutrons (35–38). The very high cross section of 4016 b (1 b = 10^{-24} cm²) for thermal neutrons makes ^{10}B essentially 100% efficient for thermal neutrons. When the ^{10}B nucleus absorbs the neutron the following processes occur (see Fig. V-40): The alpha particles (^4He) carry off most of the energy in the form of kinetic energy and are detected ' y the liquid scintillator solution the same as for any alpha particle. In 7% of the neutron absorptions, a 2.8-MeV alpha particle is produced, and in 93% of the absorptions, a 2.3-MeV alpha particle is produced. The difference in energy between the two pathways appears in the form of excitation energy of the ^7Li atom. This energy is subsequently released in the form of a γ ray of

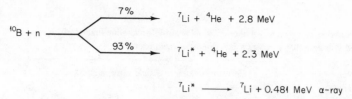

Fig. V-40. Reactions occurring when a ^{10}B nucleus absorbs a thermal neutron.

energy 0.481 MeV. Figure V-41 shows a typical spectrum of pulses obtained from a solution that was loaded with ^{10}B and irradiated with thermal neutrons. The higher energy pulses are due to the 0.481-MeV γ rays which produce Compton scattered electrons in the liquid scintillator solution.

The counting efficiency of the scintillator solution for neutrons will depend on the volume of the solution, the energy of the neutrons, and the γ-ray background. Any neutron which is absorbed by the ^{10}B will produce either a 2.3- or 2.8-MeV alpha particle which will be measured with 100% efficiency. Therefore, the counting efficiency will be given by the capture probability of the scintillator solution, concentration of ^{10}B, path length for neutrons, volume of solution, neutron capture cross section at the neutron energy, etc. Table V-9 lists the capture probabilities at different neutron energies and thickness of the scintillator solution which was loaded with 50% v/v methyl borate (95% ^{10}B) (35, 36).

A major problem in counting thermal neutrons with loaded liquid scintillator solutions is the scintillation yield for the energy released. Because of the large amounts of loaded material there is considerable quenching. Table V-10 lists the comparative scintillation yield (RPH) for various solu-

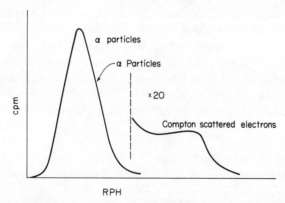

Fig. V-41. Pulse height distribution for thermal neutron detection by ^{10}B-loaded liquid scintillator. Long tail at high pulse height is due to the Compton scattered electrons produced by the 481-keV γ ray from Li*.

Table V-9

Capture probability of different volumes of liquid scintillator solution loaded with 50% v/v methyl borate (95% ^{10}B) as a function of neutron energy

Neutron energy (eV)	Capture probability for given thickness (cm)			
	1	2	2.5	5
10	0.673	0.85	0.90	0.965
173	0.28	0.50	0.60	0.86
2000	0.12	0.36	0.43	0.69
40,000	0.036	0.20	0.31	0.62

Table V-10

Relative scintillation yield of several solutions used for counting thermal neutrons

Solvent		Solutes (conc.)^a	Comparative pulse height
50%	50%		
Phenylcyclohexane–methyl borate		p-Terphenyl (4 g/liter) + DPHT (8 mg/liter)	0.63
Phenylcyclohexane–methyl borate		PPO (4 g/liter) + DPHT (16 mg/liter)	0.82
	Toluene–methyl borate	PPO (4 g/liter) + DPHT (16 mg/liter)	0.82
	Toluene–methyl borate	PPO (4 g/liter) + POPOP (20 mg/liter)	0.95
	Toluene–methyl borate	Purified PPO (4 g/liter) + purified POPOP (20 mg/liter)	1.00
	Toluene–methyl borate	PBD (4 g/liter) + POPOP (20 mg/liter)	1.06

^{a}DPHT = diphenylhexatriene.

tions. The best solution gave a yield for the alpha peak equal only to about a 44-keV electron. This low scintillation yield makes the measurement of thermal levels in the presence of significant amounts of γ rays difficult.

The low scintillation yield for the neutron-produced reactions and the production of accompanying γ rays have lead to the investigations of other ways to monitor thermal neutron levels. One such system is based on the thermal neutron capture by the isotope ^{6}Li, giving the reaction (39):

$$^6\text{Li} + \text{n} \longrightarrow {}^3\text{H} + {}^4\text{He} + 4.8 \text{ MeV}.$$

The energy released in this reaction (4.8 MeV) is almost twice that released in the ^{10}B absorption reaction (2.8 MeV). The energy is divided between the kinetic energy of the triton (^{3}H) and the alpha particle (^{4}He). Both of these particles produce excited molecules in the scintillator solution which lead to the scintillations which are counted.

Using lithium enriched in the isotope ^{6}Li (95.6%) the compound lithium salicylate has been dissolved in a scintillator solution to give about a 25-fold increase in scintillation yield per neutron compared to conventional ^{10}B-loaded solutions. The lithium salicylate is unique as it acts both as a neutron absorber and as a scintillator solute. Two solutions were employed which had the compositions (39) shown in Table V-11. Solution I differed

Table V-11

Solution compositions

	Solution I	Solution II
Dioxane	20 ml	20 ml
Li-6-salicylate	2 g	2 g
Naphthalene	2 g	2 g
Water	1 ml	0.5 ml
PPO	140 mg	0
M$_2$-POPOP	6 mg	0

basically from solution II in that PPO and M$_2$-POPOP were not present in solution II. The neutron counting efficiency is essentially 100% in both solutions. If the concentration of lithium salicylate is reduced to 1 g, the efficiency is reduced to about 60%.

Counting Fission Events

Many nuclides undergo fission, some spontaneously and some upon being excited by absorption of energy (neutrons, protons, γ rays, etc.). These nuclides can be counted in a liquid scintillator solution if dissolved in solution. To prove the counting of fission events the spontaneous fissioning nuclide ^{252}Cf was dissolved in a liquid scintillator solution as a complex of an organic phosphoric acid (5). Figure V-42 shows the distribution of pulses obtained from such a scintillator solution. The ^{252}Cf undergoes decay by two modes—97% of the decays are by emission of 6.1-MeV alpha particles and 3% of the decays are by spontaneous fission with release of 180 MeV. The alpha particles and the fission events are counted with 100% efficiencies. The 180-MeV fission events produce only as many excited molecules as an electron with only 1.3% of its energy, i.e., 2.35 MeV. Other methods (40) have determined the ratio of alpha decays to fission events (alpha/fission) to be 31.2 ± 0.3. The liquid scintillator solution data (5) gave an alpha/fission ratio of 31.0 ± 0.5.

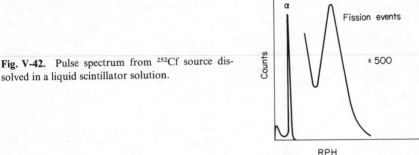

Fig. V-42. Pulse spectrum from ^{252}Cf source dissolved in a liquid scintillator solution.

Other nuclides could be determined by producing fissions in a sample dissolved in a liquid scintillator solution.

Example. What would be the fission rate for 0.1 mg uranium enriched to 1% ^{235}U, if the thermal neutron flux f were 10^6 neutrons/sec?

The weight of ^{235}U is:

$$(0.1 \text{ mg U})(0.01 \text{ fraction of } ^{235}\text{U}) = 10^{-6} \text{ g}.$$

The number of ^{235}U atoms is:

$$(6.02 \times 10^{23} \text{ atom/mole})\left(\frac{10^{-6} \text{ g}}{235 \text{ g/mole}}\right) = 2.56 \times 10^{15} \text{ atom}.$$

The fission rate R_f is defined by:

$$R_f = \sigma_f N f,$$

where σ_f is the fission cross section (5.50×10^{-22} for ^{235}U), N the number of fissionable atoms, and f the neutron flux. Therefore,

$$R_f = (5.50 \times 10^{-22})(2.56 \times 10^{15})(10^6)$$
$$= 1.4 \text{ fissions/sec}$$
$$= 84 \text{ fissions/min}.$$

Nuclear Applications

Disintegration Rate Determination

Alpha particles can be counted with 100% efficiency in most liquid scintillator solutions provided the quench level is not too great and the sample is either dissolved or finely dispersed in the scintillator solution

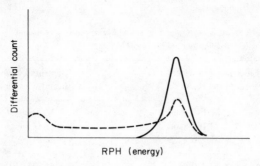

Fig. V-43. Differential pulse spectrum for alpha-particle sources dissolved (—) and partly dissolved (-----) in a liquid scintillator.

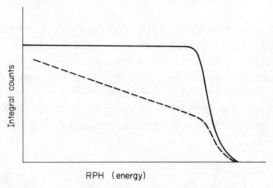

Fig. V-44. Integral pulse spectra for alpha-particle sources dissolved (—) and partly dissolved (----) in a liquid scintillator.

(24, 25, 41–43). The monoenergetic alpha particles will produce a peaked distribution of pulses as shown in Fig. V-43. If the sample is partly plated on the walls of the container or is present in a second phase, the distribution will show a tailing of pulses toward zero pulse height as shown in Fig. V-43. Samples which show this tailing may have a counting efficiency less than 100%. The same effect is reflected in the integral pulse counts as shown in Fig. V-44. The dissolved sample shows no increase in counts below the peak area, whereas the partly dissolved source has an increasing count over the whole pulse height span.

The counting efficiency of a liquid scintillation counter was compared to an intermediate geometry alpha counter (IGAC) (24). Equal aliquots of a stock solution of ^{239}Pu were obtained. One aliquot was dissolved in a liquid scintillator and counted in a liquid scintillation counter, while the other aliquot was plated on a platinum disk and counted in the IGAC. The

Table V-12

Comparison of alpha count rate obtained in a liquid scintillator system and
the disintegration rate obtained in a calibrated IGAC[a]

Efficiency-converted dpm in the IGAC[b]	Measured cpm in liquid scintillator
$31,142 \pm 58$	$32,293 \pm 56$
$32,164 \pm 58$	$31,833 \pm 56$
$31,956 \pm 58$	$32,187 \pm 25$
Average $32,090 \pm 51$	$32,104 \pm 45$

[a] See P. M. Wright, E. P. Steinberg, and L. E. Glendenin, *Phys. Rev.* **123,** 205 (1961).
[b] Efficiency of IGAC was 0.09633 ± 0.00001.

results are compared in Table V-12. The dpm calculated for the sample
counted in the IGAC was well within statistical agreement with the measured cpm in the liquid scintillation counter. Thus the liquid scintillator
provides 4π geometry (100% efficiency) for the alpha particles.

The disintegration rate of a ^{239}Pu sample was obtained from several
different methods (43), the results of which are listed in Table V-13. All of
the values agree with the standard value obtained by the National Physical
Laboratory (England).

Table V-13

Standardization of ^{239}Pu source alpha specific activity by several different methods[a]

Method	Counting data technique	$(dpm/g) \times 10^{-4}$
Sample dissolved in LSS[b]	Integral count	664 ± 5
Sample on filter paper in LSS	Integral count	681 ± 4
Sample dissolved in LSS	Differential count, bracketing the peaks	660 ± 9
Plated between two plastic disks	Integral count	668 ± 3
Low geometry proportional counter	Integral count	668 ± 3
Standard value (NPL)[c]		671 ± 4

[a] See G. A. Brinkman, Ph.D. Thesis, Univ. of Amsterdam, 1961.
[b] LSS = liquid scintillation solution.
[c] NPL = National Physics Laboratory, England.

The determination of the disintegration rates of radionuclides that decay
by beta emission (negatrons and positrons) is more difficult because the
energy distribution spans from some maximum energy E_{max} to zero energy.
No matter how great E_{max} is there are always some beta particles which
have insufficient energy to produce a measurable pulse in the liquid scintil-

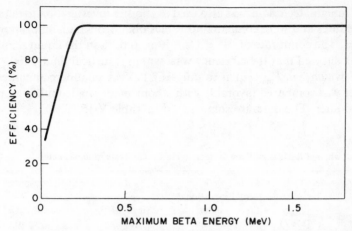

Fig. V-45. Expected counting efficiency as a function of the maximum beta energy for the radioactive beta emission of the nuclide.

lation counter. A series of beta emitters of different E_{max} was counted in a commercial liquid scintillation counter (44). The measured counting efficiency is plotted in Fig. V-45 as a function of E_{max} of the beta transition. Above an E_{max} of 200 keV the integral counting efficiency is essentially 100%.

Using the technique of integral extrapolation to zero pulse height, the count rates of ^{131}I, ^{32}P, and ^{60}Co were measured in a liquid scintillator solution (45, 46). The count rates obtained with the liquid scintillator were well within the limits obtained by the National Physical Laboratory for standardized solutions of these radionuclides. The data are summarized in Table V-14.

Table V-14

Comparison of disintegration rates determined by the integral count method in liquid scintillator systems and other methods[a]

Isotope	Obtained by	μCi/g
^{131}I	Liquid scintillation	25.19
	National Physical Laboratory standard	$25.0 \pm 5\%$
^{32}P	Liquid scintillation	23.50
	4π proportional	$23.60 \pm 1\%$
^{60}Co	Liquid scintillation	10.42
	4π Geiger–Meuller	$10.40 \pm 3\%$
	$\beta-\gamma$ coincidence	$10.74 \pm 2\%$

[a] See J. Steyn, *Proc. Phys. Soc. London Sect. A* **69**, 865 (1956).

The integral technique was also used in another investigation to determine sample rates and were compared to values obtained with a 4π proportional counter. The count rate of ^{32}P, ^{137}Cs, ^{147}Pm, and ^{35}S in the liquid scintillator counter showed that the efficiency was, within statistical limits, 100%, even for ^{35}S which has E_{max} equal to only 167 keV. A commercial coincidence counter was compared favorably with a homemade single multiplier phototube counter. The data are summarized in Table V-15.

Table V-15

Standardization of liquid scintillator systems against a $4\pi\beta$ proportional counter[a]

Nuclide	E_{max} (MeV)	$4\pi\beta$ Proportional counter cpm = dpm	Liquid scintillation counter	
			Single MPT (cpm)	Coincidence 2 MPT (cpm)
^{32}P	1.707	47,400	46,800	47,200
^{137}Cs	0.5	40,500	39,400	41,200
^{147}Pm	0.22	125,500	126,000	126,000
^{35}S	0.167	64,880	64,500	65,000

[a] See K. F. Flynn and L. E. Glendenin, *Phys. Rev.* **116**, 744 (1959).

Many other investigations have been performed to show the counting efficiency of liquid scintillation counters for different radionuclides and different types of samples. Some of these are given in Table V-16.

Table V-16

Some published results for the determination of disintegration rates of beta emitters

Nuclides	Techniques	Results	Ref.
^{24}Na, ^{32}P, ^{42}K, ^{89}Sr, ^{90}Sr, ^{131}I, ^{198}Au	1. Sample deposited between two thin cellulose acetate sheets and submerged in liquid scintillator solution. 2. Integral count with varying foil thickness. Extrapolation to zero foil thickness	Considerable error, $\sim 10\%$, due to two extrapolations. Higher errors for lower energy β emitters, ^{60}Co ($E_{max} = 0.31$ MeV) 30–40% errors	a
^{131}I, ^{32}P, ^{60}Co	1. Dissolved sample 2. Integral counting	Approximately 1% agreement with $4\pi\beta$ proportional and $4\pi\beta$–γ coincidence counter (^{60}Co)	b

Table V-16 (continued)

Nuclides	Techniques	Results	Ref.
^{32}P, ^{24}Na	1. Dissolved sample or production of radioactive nuclide in liquid scintillator solution by neutron irradiation 2. Integral counting	Not given but claimed to be good	c
^{32}P, ^{137}Cs, ^{147}Pm, ^{35}S, ^{87}Rb, ^{45}Ca	1. Dissolved samples 2. Integral counting	Conservative estimate of overall reliability $\sim 2\%$, compared to $4\pi\beta$ proportional counter. Measured half-life of ^{87}Rb	d
^{3}H, ^{241}Pu, ^{106}Ru–^{106}Rh, ^{63}Ni, ^{151}Sm, ^{14}C, ^{35}S	1. Dissolved samples 2. Integral counting along with a theoretically calculated maximum counting efficiency	Good agreement between experimental and theoretical data. Where standards were available 2% agreement	e
^{63}Ni	1. Dissolved sample 2. Integral counting and theoretical counting efficiency	Obtained ^{63}Ni half-life	f
^{40}K	1. Gel suspension of K_2SO_4 2. Integral counting	Obtained ^{40}K half-life	g
^{90}Sr–^{90}Y, ^{32}P, ^{24}Na ^{198}Au, ^{131}I, ^{137}Cs, ^{60}Co, ^{35}S, ^{14}C, ^{40}K	1. Dissolved, suspended, and filter paper samples 2. Integral counting	Dissolved samples higher energies $\sim 98\%$, lower energies $\sim 90\%$ Filter paper samples higher energies $\sim 95\%$, lower energies $< 80\%$	h
^{137}Cs	1. Dissolved sample 2. Integral counting	Obtained ^{137}Cs half-life	i
^{147}Pm	1. Dissolved sample 2. Coincidence counting system with fixed energy window	Usually 5% higher than $4\pi\beta$ proportional counter	j
Several with E_{max} from 17 keV to 1.7 MeV	1. Dissolved samples 2. Integral counting	100% efficiencies for E_{max} > 200 keV	k

Table V-16 (continued)

Nuclides	Techniques	Results	Ref.
^{85}Kr	1. Kr gas dissolved in liquid scintillator solvent (toluene) 2. Integral counting	Determined solubility of Kr in toluene	[l]
^{63}Ni, ^{14}C, ^{35}S	1. Dissolved samples 2. Integral counting of quenched samples. Extrapolation of integral counting rates to zero quencher	100% efficiencies within 2%	[m]
^{3}H, ^{63}Ni, ^{14}C, ^{95}Nb	1. Dissolved samples 2. Integral counting with light filters. Extrapolation of integral counting rates to zero figure of merit	100% efficiencies within 2%	[n]

[a] E. H. Belcher, *J. Sci. Instrum.* **30**, 286 (1953).

[b] J. Steyn, *Proc. Phys. Soc. London Sect. A* **69**, 865 (1956).

[c] E. T. Józefowicz, *Nukleonika* **5**, 713 (1960).

[d] K. F. Flynn and L. E. Glendenin, *Phys. Rev.* **116**, 744 (1959).

[e] D. L. Horrocks and M. H. Studier, *Anal. Chem.* **33**, 615 (1961).

[f] D. L. Horrocks and A. L. Harkness, *Phys. Rev.* **125**, 1619 (1962).

[g] D. G. Fleishman and V. V. Glazunov, *Sov. J. At. Energy* **12**, 338 (1962); *At. Energ.* **12**, 320 (1962).

[h] G. A. Brinkman, Ph.D. Thesis, Univ. of Amsterdam, 1961.

[i] D. G. Fleishman, I. V. Burovina and V. P. Nesterov, *Sov. J. At. Energy* **13**, 1224 (1963); *At. Energ.* **13**, 592 (1962).

[j] J. D. Ludwick, *Anal. Chem.* **36**, 1104 (1964).

[k] G. Goldstein, Chemistry Division-Progress Report Period Ending June 20, 1964. Rep. ORNL 3679-UC-4-TID-4500, p. 19. Oak Ridge Nat. Lab., Oak Ridge, Tennessee, 1964.

[l] D. L. Horrocks and M. H. Studier, *Anal. Chem.* **36**, 2077 (1964).

[m] D. L. Horrocks, *Prog. Nucl. Energy Ser. 9* **7**, 21–110 (1966).

[n] K. F. Flynn, L. E. Glendenin and V. Prodi, *in* "Organic Scintillators and Liquid Scintillation Counting" (D. L. Horrocks and C. T. Peng, eds.), pp. 687–696. Academic Press, New York, 1971.

However, the major use of liquid scintillation counters is for the measurement of beta emitters of E_{max} energy less than 200 keV. These low energies do not produce a counting efficiency of 100%. The counting efficiency is usually determined by the use of samples that have been standardized by other methods. The National Bureau of Standards (Rockville, Maryland) has been instrumental in preparing many of these standardized samples.

Several theoretical methods have been published for calculation of the counting efficiency of a liquid scintillation counter for beta emitters of E_{max} less than 200 keV (47–50). Assuming an average energy required to produce a measurable pulse, called the figure of merit, it is possible to calculate the counting efficiency of a radionuclide from its known beta-energy distribution. Table V-17 lists some calculated efficiencies for a single multiplier phototube counter and a commercial coincidence counter at different figure of merit values.

Table V-17

Theoretically calculated counting efficiencies for different radionuclides as a function of the figure of merit

| Isotope | E_{max} (keV) | Figure of merit (keV) | | | | | |
| | | 1.0 | | 1.5 | | 2.0 | |
		Single	Coincidence	Single	Coincidence	Single	Coincidence
^{3}H	18	0.91	0.71	0.86	0.59	0.82	0.50
^{241}Pu	21	0.87	0.65	0.82	0.54	0.77	0.44
^{106}Ru	39	0.94	0.81	0.91	0.73	0.88	0.65
^{63}Ni	67	0.96	0.87	0.93	0.81	0.91	0.75
^{151}Sm	76	0.97	0.90	0.95	0.85	0.93	0.80
^{14}C	156	0.991	0.971	0.987	0.956	0.981	0.939
^{35}S	167	0.988	0.962	0.982	0.943	0.975	0.923

The value of the figure of merit can be obtained by determining the ratio of count rates from a single multiplier phototube and a coincidence pair of multiplier phototubes (one tube common to both measurements). The ratio will be unique for each radionuclide. Counting several radionuclides will give several ratios which should correspond to one figure of merit. The theoretical calculations of the counting efficiencies and ratios for several radionuclides are shown in Figs. V-46–V-49.

Since the figure of merit (k) is the energy expended in the solution to produce a measurable pulse, this will be proportional to the pulse height of the Compton edge of a γ-ray source. If the value of k doubles the pulse height corresponding to the Compton edge of a given γ-ray energy would be halved. Thus the relationship can be expressed:

$$k_q = k_{qo} \frac{(\text{RPH})_{qo}}{(\text{RPH})_q}$$

where k_{qo} is the figure of merit for a least quenched solution, k_q the figure of merit for a quenched solution, $(\text{RPH})_{qo}$ the relative pulse height for the

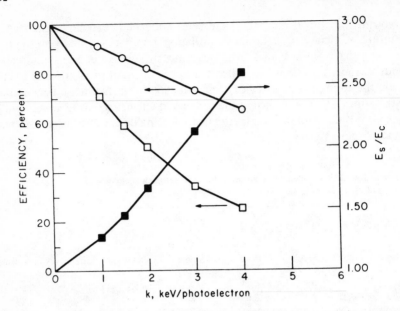

Fig. V-46.   Counting efficiencies in single channel E_s (○) and coincidence E_c (□) type counting systems and E_s/E_c ratio (■) for ^{3}H as function of figure of merit k.

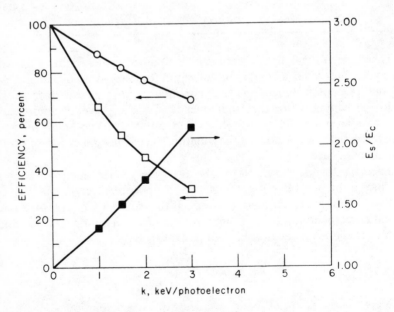

Fig. V-47.   Counting efficiencies in single channel E_s (○) and coincidence E_c (□) type counting systems and E_s/E_c ratio (■) for ^{241}Pu as function of figure of merit k.

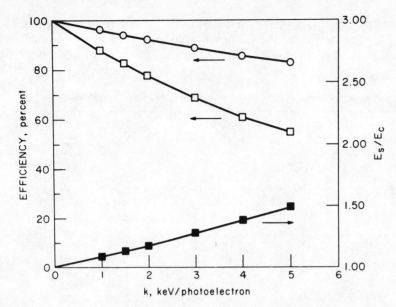

Fig. V-48. Counting efficiencies in single channel E_s (○) and coincidence E_c (□) type counting systems and E_s/E_c ratio (■) for ^{63}Ni as function of figure of merit k.

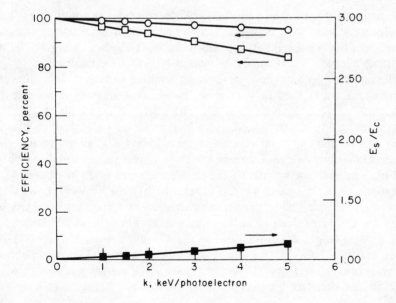

Fig. V-49. Counting efficiencies in single channel E_s (○) and coincidence E_c type (□) counting systems and E_s/E_c ratio (■) for ^{14}C as function of figure of merit k.

Compton edge of given γ ray on a least quenched solution, and $(RPH)_q$ the relative pulse height for the Compton edge of given γ ray on a quenched solution. Table V-18 shows a comparison of the value of k obtained by

Table V-18

Figures of merit as function of quench level determined by two different methods

Radionuclide	μl EtOH $\overline{250\ \mu l\ \text{LSS}}$	$\dfrac{\int \text{cpm by single}}{\int \text{cpm by coincidence}}$	$\dfrac{(RPH)_{q_0}}{(RPH)_q}$	k_{cpm}	k_q
^{63}Ni	0	1.058	1.000	0.65	0.65^a
	25	1.105	1.725	1.15	1.12
	50	1.102	1.771	1.13	1.15
	75	1.125	2.161	1.38	1.41
	100	1.148	2.570	1.60	1.67
^{14}C	0	1.015	1.000	0.71	0.71^a
	25	1.020	1.402	1.00	1.02
	50	1.026	1.742	1.24	1.26
	75	1.032	2.078	1.48	1.49
	100	1.038	2.480	1.76	1.78

$^a k_q$ at zero quench is taken equal to k_{cpm} at zero quench since k_q determinations are only relative numbers.

two methods, quench and ratio of single to coincidence count rates. The agreement is quite good between the two methods for a given radionuclide at different quench levels; the agreement between the two radionuclides at the same quench level is somewhat poorer but still acceptable.

The relationship between k_q at different levels of quench and the integral counting rate was used to determine the absolute disintegration rate of quenched samples. By definition k is the energy required to produce a measurable pulse. If k were zero, every beta particle would produce a measurable pulse and the counting efficiency would be 100%. By measuring the integral counting rate of a sample which has been successively quenched and plotting that rate against k_q, the disintegration rate can be obtained by extrapolation of the rate to k equal to zero (50, 51). Figure V-50 shows such a plot for a sample ^{14}C of known disintegration rate. It is not necessary to know the absolute value of k at any given quench. Only the value of the ratio of the pulse height corresponding to the Compton edge of the 662-keV γ rays of the ^{137}Cs–^{137m}Ba source need be measured. The extrapolated integral counting rate at a ratio of zero gave 100% counting efficiency with an estimated error well within the $\pm 1.5\%$ uncertainty of the standard ^{14}C sample.

The major drawback to this technique is that the sample has been quenched by the addition of a quenching agent to the scintillator solution, and the

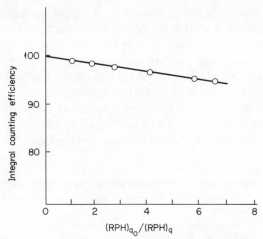

Fig. V-50. Plot of integral count rate of ^{14}C sample as function of quench as measured by Compton edge (RPH) showing extrapolation to 100% counting efficiency.

sample cannot be recovered in its unquenched form for future counting or rechecking of its counting rate.

To eliminate this drawback a technique involving the use of external standard absorption filters has been utilized with considerable success (52). Figure V-51 shows the apparatus and the relative position of the sample and the filters. For the same measurements to be made in a coincidence-type liquid scintillation counter identical filters would have to be placed in front of each of the two multiplier phototubes. This is almost impossible with the commercially available coincidence liquid scintillation counters.

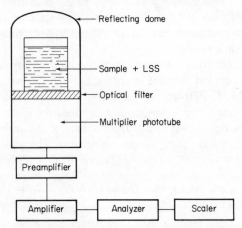

Fig. V-51. Simple counting apparatus used with various optical filters to simulate quench in a method for obtaining disintegration rate of sample.

The optical filters absorb a fraction of the photons emitted from the scintillator and, in effect, change the threshold of detection. This is essentially the same as increasing the value of k. If the optical density of the filters is known for the wavelength band of emission from the scintillator solutes, the integral count rate can be plotted against the optical density ratio. The disintegration rate is obtained by extrapolation to a ratio value of zero. However, it is not necessary to know the optical density of each filter. The change in detection level can be monitored by the change in the pulse height of the Compton edge of an external γ-ray source. Table V-19 lists

Table V-19

Comparison of disintegration rates obtained by the optical filter technique and other methods[a]

Radionuclide	Maximum beta energy (keV)	dpm by filter method	Standardized dpm	Method of standardizing
^{3}H	18	$141,000 \pm 2\%$	$142,500 \pm 2\%$	Ionization counting
^{106}Ru	40	$71,000 \pm 3\%$	$72,200 \pm 2\%$	Equilibrium with radioactive daughter
^{14}C	158	$150,000 \pm 2\%$	$148,000 \pm 2\%$	Absolute beta counting
^{95}Nb	160	$167,000 \pm 2\%$	$169,000 \pm 2\%$	$4\pi\beta-\gamma$ coincidence
^{60}Co	314	$130,200 \pm 2\%$	$130,800 \pm 2\%$	$4\pi\beta-\gamma$ coincidence

[a] See K. F. Flynn, L. E. Glendenin, and V. Prodi, *in* "Organic Scintillators and Liquid Scintillation Counting" (D. L. Horrocks and C. T. Peng, eds.), p. 687. Academic Press, New York and London, 1971.

several low-energy beta-emitting nuclides and the measured integral count rate by the optical filter method. The actual disintegration rate, determined by other methods, is also listed. In all cases the agreement is within the limits of the determinations.

Monoenergetic electrons from certain radionuclides can be counted with 100% efficiency provided they are energetic enough to exceed the threshold of detection (3, 17, 53). However, since most radionuclides which produce monoenergetic electrons by decay do not produce the given monoenergetic electron for every decay, the disintegration rate of the radionuclide is not equal to the rate of the monoenergetic electrons. The rate of emission of monoenergetic electrons is usually obtained from the decay scheme for the radionuclide. Consider the decay of ^{131m}Xe, which is schematically represented (13) in Fig. V-52. The 164-keV transaction is split between the emission of a 164-keV γ ray and conversion electrons resulting from K and L shells and the subsequent X rays and Auger electrons. The γ rays have essentially a zero probability of depositing their total energy in the liquid

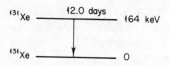

Fig. V-52. Schematic of the decay of ¹³¹ᵐXe.

scintillator. The few that do interact with the scintillator medium do so by production of Compton scattered electrons that have less than 65-keV energy. The conversion electrons have a 100% probability of being stopped and giving up their total energy to the liquid scintillator solution. Thus knowing the ratio between electron emission and gamma emission, the disintegration rate could be calculated from the rate of conversion electrons.

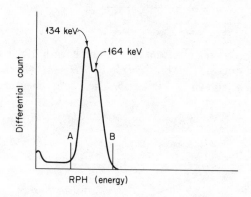

Fig. V-53. Pulse spectrum for sample of ¹³¹ᵐXe dissolved in a liquid scintillator solution–toluene solvent.

Figure V-53 shows the spectrum of counts obtained from a sample of ¹³¹ᵐXe in a liquid scintillator. The sum of counts that occur within the pulse height limits of A and B is the total due to the conversion electrons. The ratio of electron to gamma emission (e/γ) is 41. This corresponds to 41/42, or 97.6% of the ¹³¹ᵐXe decays occurring by emission of conversion electrons. Thus the true decay rate of ¹³¹ᵐXe is given by

$$^{131m}\text{Xe dpm} = \left(\int \text{conversion electron cpm} \right) \Big/ 0.976.$$

The double-peaked distribution of Fig. V-53 is actually due to conversions occurring in the K shell (134 keV) and in the L shell (164 keV).

A somewhat more complicated decay scheme, which includes the emission of conversion electrons in a part of the decays, is that for the radionuclide pair ¹³⁷Cs–¹³⁷ᵐBa. The ¹³⁷Cs decays by the emission of beta particles;

7% (E_{max} = 1.17 MeV) leads directly to the ground state of ^{137}Ba, and 93% (E_{max} = 0.52 MeV) to the metastable excited state ^{137m}Ba. The ^{137m}Ba decays with an independent half-life (2.6 min) by emission of γ rays and conversion electrons with associated X rays and Auger electrons. The decay scheme (13) is shown diagrammatically in Fig. V-54.

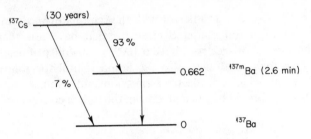

Fig. V-54. Decay scheme of ^{137}Cs–^{137m}Ba.

Again the conversion electrons are counted with 100% efficiency. The count spectrum obtained with a sample of ^{137}Cs–^{137m}Ba in a liquid scintillator solution is shown in Fig. V-55. The counts measured in windows

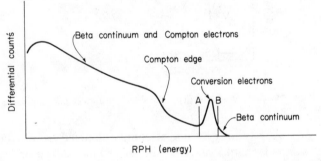

Fig. V-55. Pulse spectrum of sample of ^{137}Cs–^{137m}Ba (as salt of octoic acid) dissolved in a liquid scintillator.

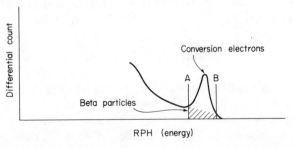

Fig. V-56. Method for obtaining conversion electron count rate by subtraction of pulses due to beta continuum.

A to B (Fig. V-55) are due to conversion electrons of ^{137m}Ba and the beta particles from the 7% beta branch ($E_{max} = 1.18$ MeV) of ^{137}Cs. The beta continuum is a smooth function of counts versus relative pulse height. The "background" due to the beta continuum can be obtained by the extrapolation method demonstrated in Fig. V-56.

The e/γ ratio for the 662-keV transition is 0.11, or 0.11/1.11 of ^{137m}Ba decays produce conversion electrons. Thus the decay rate of ^{137m}Ba as obtained from the corrected conversion electron count rate is

$$^{137m}\text{Ba dpm} = \left(\int \text{conversion electron cpm} \right) \Big/ 0.10.$$

From the ^{137m}Ba dpm value the decay rate of ^{137}Cs is

$$^{137}\text{Cs dpm} = (^{137m}\text{Ba dpm})/0.93.$$

Determination of Half-Lives

The absolute counting obtainable in the liquid scintillator solution makes it ideal for determining the half-lives of radionuclides by the specific activity method. If a known number of atoms of a given nuclide (N) is incorporated into a liquid scintillator solution and the disintegration rate of that radionuclide measured, the half-life can be calculated from the relationship:

$$t_{1/2} \text{ (min)} = \frac{(0.693)N}{\text{dpm}}.$$

The ability to count low-energy beta emitters has made it easier to determine the half-life of several radionuclides of short to intermediate half-life by following the decrease in activity with time. The higher counting efficiency of the liquid scintillator solutions leads to more statistically significant count rate determinations.

The half-lives of several radionuclides have been determined using the liquid scintillation system. The radionuclides and their measured half-lives are given in Table V-20. In some cases the values obtained substantiated the accepted half-life and reduced the known error. In other cases new half-lives were actually measured.

Determination of Beta Endpoint Energies

The energy response of a scintillator solution for electrons is known (see pp. 90–92). From this relationship it is possible to use the pulse height spectrum obtained from a liquid scintillator solution to obtain an

Table V-20

Half-life values measured by liquid scintillation methods

Nuclide	Type of particle, energy (MeV)	Technique	Measured half-life	Ref.
^{63}Ni	β^-, $E_{max} = 0.067$	Specific activity	91.6 ± 3.1 years	a
^{147}Sm	α, 2.2	Specific activity	$(1.05 \pm 0.02) \times 10^{11}$ years	b
^{87}Rb	β^-, $E_{max} = 0.272$	Specific activity	$(47.0 \pm 1.0) \times 10^9$ years	c
^{137}Cs	β^-, $E_{max} = 1.2, 0.51$	Specific activity	30.1 ± 0.7 years	d
^{232}Th	α, 4.0	Specific activity	$(1.40 \pm 0.007) \times 10^{10}$ years	e
^{238}U	α, 4.2	Specific activity	$(4.46 \pm 0.01) \times 10^9$ years	f
^{211}At	α, 5.8	Decay	7.3 h	g
^{35}S	β^-, $E_{max} = 0.167$	Decay	87.17 ± 0.03 days	h
^{45}Ca	β^-, $E_{max} = 0.255$	Decay	163 ± 1 days	h
^{40}K	β^- (89%), $E_{max} = 1.33$ E.C. (11%)	Specific activity	$(1.48 \pm 0.02) \times 10^9$ years	i
^{32}Si	β^-, $E_{max} = 0.213$	Log ft	~ 200 years	j
^{36}Cl	β^- (98.3%), $E_{max} = 0.714$ E.C. (1.7%)	Specific activity	$(3.06 \pm 0.02) \times 10^5$ years	k
^{10}Be	β^-, $E_{max} = 0.555$	Specific activity	$(1.91 \pm 0.15) \times 10^6$ years	k
^{99}Tc	β^-, $E_{max} = 0.290$	Specific activity	2.1×10^5 years	l

a D. L. Horrocks and A. L. Harkness, *Phys. Rev.* **125**, 1619 (1962).

b P. M. Wright, E. P. Steinberg, and L. E. Glendenin, *Phys. Rev.* **123**, 205 (1961); G. B. Beard and W. H. Kelly, *Nucl. Phys.* **8**, 207 (1958); D. Donhoffer, *Nucl. Phys.* **50**, 489 (1964).

c K. F. Flynn and L. E. Glendenin, *Phys. Rev.* **116**, 744 (1960).

d D. G. Fleishman, I. V. Burovina, and V. P. Nesterov, *Sov. J. At. Energy* **13**, 1224 (1963); *At. Energ.* **13**, 592 (1962).

e L. J. LeRoux and L. E. Glendenin, *Nat. Conf. Nucl. Energy, Appl. Isotop. and Radiat.* (F. L. Warren, ed.), pp. 83–94. At. Energy Board, Petoria, South Africa, 1963.

f J. Steyn and F. W. Strelow, *Metrol. of Nuclides Symp., October 1959*, pp. 155–161. IAEA, Vienna, 1960.

g J. K. Basson, *Anal. Chem.* **28**, 1472 (1956).

h R. Vaninbroukx and A. Spernol, *Int. J. Appl. Radiat. Isotop.* **16**, 289 (1965).

i D. G. Fleishman and V. V. Glazunov, *Sov. J. At. Energy* **12**, 338 (1962); *At. Energ.* **12**, 320 (1962).

j R. L. Brodzinski, J. R. Finkel and D. C. Conway, *J. Inorg. Nucl. Chem.* **26**, 677 (1964).

k G. Goldstein, Annu. Progr. Rep., p. 50. Anal. Chem. Div., Oak Ridge Nat. Lab. Oak Ridge, Tennessee, 1966.

l G. Goldstein, Annu. Progr. Rep., ORNL-3750, pp. 51–53. Anal. Chem. Div., Oak Ridge Nat. Lab., Oak Ridge, Tennessee, 1964.

energy distribution of beta particles from the given radionuclide. The energy distributions have been used to obtain Kurie plots. In some cases a second-derivative correction formula has been shown necessary to correct for energy resolution of the liquid scintillator. Table V-21 lists several beta (negatron and positron) emitters that have been analyzed in the liquid

Table V-21

Some E_{max} values determined by liquid scintillation techniques

Nuclide	Mode of decay	Accepted[a] E_{max} (keV)	LSS measured E_{max} (keV)	Ref.
^{87}Rb	β^-	275	272 ± 3	b
^{22}Na	β^+	541	533 ± 7	c
^{45}Ca	β^-	254	253 ± 3	b
^{63}Ni	β^-	67	67 ± 2	d
^{32}Si	β^-	210	213 ± 7	e
^{14}C	β^-	156	154 ± 5	e
^{32}P	β^-	1707	1690 ± 30	f

[a] C. M. Lederer, J. M. Hollander, and I. Perlman, "Table of Isotopes," 6th ed. Wiley, New York, 1967.
[b] K. F. Flynn and L. E. Glendenin, *Phys. Rev.* **116**, 744 (1960).
[c] D. W. Engelkemier, K. F. Flynn, and L. E. Glendenin, *Phys. Rev.* **126**, 1818 (1962).
[d] D. L. Horrocks and A. L. Harkness, *Phys. Rev.* **125**, 1619 (1962).
[e] R. L. Brodzinski, J. R. Finkel, and D. C. Conway, *J. Inorg. Nucl. Chem.* **26**, 677 (1964).
[f] T. Cless-Bernert, K. E. Duftschmid, and H. Vonach, *Nukleonik* **4**, 162 (1962).

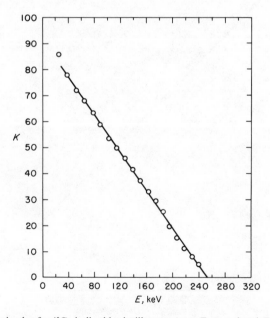

Fig. V-57. Kurie plot for ^{45}Ca in liquid scintillator system. Extrapolated E_{max} is 253 keV.

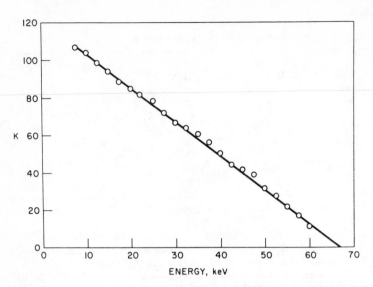

Fig. V-58.   Kurie plot for ^{63}Ni in liquid scintillator system. Extrapolated E_{max} is 67 keV.

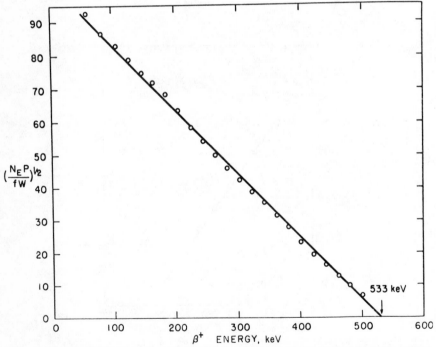

Fig. V-59.   Kurie plot for ^{22}Na (positron emitter) in liquid scintillator system. Extrapolated E_{max} is 533 keV.

scintillator system and the E_{max} values obtained by the Kurie plots. Figure V-57 shows the Kurie plot for a ^{45}Ca source in a liquid scintillator. The extrapolated energy value at $K = 0$ is the value of E_{max}. The E_{max} obtained from the plot of Fig. V-57 is 253 keV, which agrees with the accepted value of 254 keV. Similar plots for obtaining the values of E_{max} for ^{63}Ni and the positron emitter ^{22}Na are shown in Figs. V-58 and V-59, respectively.

References

1. K. F. Flynn, L. E. Glendenin, E. P. Steinberg, and P. M. Wright, *Nucl. Instrum. Methods* **27,** 13 (1964).
2. D. L. Horrocks, *Rev. Sci. Instrum.* **35,** 334 (1964).
3. D. L. Horrocks, *Nucl. Instrum. Methods* **30,** 157 (1964).
4. R. Batchelor, W. B. Gilboy, J. B. Parker, and J. H. Towle, *Nucl. Instrum. Methods* **13,** 70 (1961).
5. D. L. Horrocks, *Rev. Sci. Instrum.* **34,** 1035 (1963).
6. K. F. Flynn and L. E. Glendenin, *Phys. Rev.* **116,** 744 (1959).
7. D. W. Engelkemeir, K. F. Flynn, and L. E. Glendenin, *Phys. Rev.* **126,** 1818 (1962).
8. F. N. Hayes, B. S. Rogers, P. Sanders, R. L. Schuch, and D. L. Williams, Rep. LA-1639. Los Alamos Sci. Lab., Los Alamos, New Mexico, 1953.
9. F. N. Hayes, D. G. Ott, V. N. Kerr, and B. S. Rogers, *Nucleonics* **13,** No. 12, 38 (1955).
10. F. N. Hayes, D. G. Ott, and V. N. Kerr, *Nucleonics* **14,** No. 1, 42 (1956).
11. W. J. Kaufman, A. Nir, G. Parks, and R. M. Hours, *Proc. Conf. Organic Scintillation Detectors, Univ. of New Mexico, 1960* (G. H. Daub, F. N. Hayes, and E. Sullivan, eds.) TID-7612, p. 239. U.S. At. Energy Commission, Washington, D.C., 1961.
12. W. J. Kaufman, A. Nir, G. Parks, and R. M. Hours, "Tritium in the Physical and Biological Sciences," Vol. 1, p. 249. IAEA, Vienna, 1962.
13. C. M. Lederer, J. M. Hollander, and I. Perlman, "Table of Isotopes," 6th ed. Wiley, New York, 1967.
14. D. L. Horrocks, *Int. J. Appl. Radiat. Isotop.* **22,** 258 (1971).
15. J. K. Basson and J. Steyn, *Proc. Phys. Soc. London Sect. A* **67,** 297 (1954).
16. J. K. Basson, *Anal. Chem.* **28,** 1472 (1956).
17. D. L. Horrocks and M. H. Studier, *Anal. Chem.* **36,** 2077 (1964).
18. D. L. Horrocks, *Int. J. Appl. Radiat. Isotop.* **17,** 441 (1966).
19. W. J. McDowell, *in* "Organic Scintillators and Liquid Scintillation Counting" (D. L. Horrocks and C. T. Peng, eds.) p. 937. Academic Press, New York, 1971.
20. W. D. Fairman and J. Sedlet, *Anal. Chem.* **40,** 2004 (1968).
21. D. L. Horrocks, unpublished results, 1971.
22. I. B. Berlman, R. Grismore, and B. G. Oltman, *Trans. Faraday Soc.* **59,** 2010 (1963).
23. D. L. Horrocks, *in* "Organic Scintillators" (D. L. Horrocks, ed.), p. 45. Gordon & Breach, New York, 1968.
24. P. M. Wright, E. P. Steinberg, and L. E. Gendenin, *Phys. Rev.* **123,** 205 (1961).
25. D. L. Horrocks and M. H. Studier, *Anal. Chem.* **30,** 1747 (1958).
26. H. R. Ihle, M. Karayannis, and A. P. Murrenhoff, "Radioisotope Sample Measurement Techniques in Medicine and Biology," p. 485. IAEA, Vienna, 1965.
27. A Lindenbaum and C. J. Lund, *Radiat. Res.* **37,** 131 (1969).
28. D. L. Horrocks, To be published, 1974.

29. D. L. Horrocks, *Nature (London)* **202,** 78 (1964).
30. J. Ashcroft, *Anal. Biochem.* **37,** 268 (1970).
31. A. R. Ronzio, *Int. J. Appl. Radiat. Isotop.* **4,** 196 (1959).
32. R. Batchelor, W. B. Gilboy, J. B. Parker, and J. H. Towle, *Nucl. Instrum. Methods* **13,** 70 (1961).
33. H. W. Broek and C. E. Anderson, *Rev. Sci. Instrum.* **32,** 1063 (1960).
34. G. E. Clark, R. C. Morrison, J. W. O'Laughlin, and H. R. Burkholder, *Nucl. Instrum. Methods* **84,** 67 (1970).
35. C. O. Muehlhause and G. E. Thomas, *Phys. Rev.* **85,** 926 (1952).
36. C. O. Muehlhause and G. E. Thomas, *Nucleonics* **11,** No. 1, 44 (1953).
37. L. M. Bollinger and G. E. Thomas, *Rev. Sci. Instrum.* **28,** 489 (1957).
38. G. E. Thomas, *Nucl. Instrum. Methods* **17,** 132 (1962).
39. H. H. Ross and R. E. Yerick, *Nucl. Sci. Eng.* **20,** 23 (1964).
40. D. Metta, H. Diamond, R. F. Barnes, H. Milsted, J. Gray, Jr., D. J. Henderson, and C. M. Stevens, *J. Inorg. Nucl. Chem.* **27,** 33 (1965).
41. H. H. Seliger, *Int. J. Appl. Radiat. Isotop.* **8,** 29 (1960).
42. G. B. Beard and W. H. Kelly, *Nucl. Phys.* **8,** 207 (1958).
43. G. A. Brinkman, Ph.D. Thesis, Univ. of Amsterdam, 1961.
44. G. Goldstein, *Nucleonics* **23,** No. 3, 67 (1965).
45. J. Steyn, *Proc. Phys. Soc. London Sect. A* **69,** 865 (1956).
46. J. Steyn and F. J. Haasbroeck, *Proc. U.N. Int. Conf. Peaceful Uses At. Energy, 2nd Geneva, 1958,* **21,** p. 95. U.N., New York, 1958.
47. D. L. Horrocks and M. H. Studier, *Anal. Chem.* **33,** 615 (1961).
48. H. R. Lukens, Jr., *Int. J. Appl. Radiat. Isotop.* **12,** 134 (1961).
49. J. A. B. Gibson and H. J. Gale, *J. Sci. Instrum.* [2], **1,** 99 (1968).
50. D. L. Horrocks, *Progr. Nucl. Energy Ser. 9* **7,** 21–110 (1966).
51. D. L. Horrocks, *Nature (London)* **202,** 78 (1964).
52. K. F. Flynn, L. E. Glendenin, and V. Prodi, *in* "Organic Scintillators and Liquid Scintillation Counting" (D. L. Horrocks and C. T. Peng, eds.), p. 687. Academic Press, New York, 1971.
53. D. L. Horrocks, "Developments in Applied Spectroscopy," Vol. 7A, p. 303. Plenum Press, New York, 1969.

CHAPTER VI

PREPARATION OF COUNTING SAMPLES

Sample preparation for liquid scintillation counting is a very complicated topic about which to write. There are so many types of samples and each type presents a separate problem. Of course, the most simple type of sample preparation is none; that is, the sample can be directly dissolved in the liquid scintillator solution and its activity level measured with no correction necessary (except for counting efficiency). Unfortunately, this is not the case for the majority of the sample types that are counted in liquid scintillation systems.

One general consideration for all sample preparation is the sample purity —both chemical and radioactive purity are important. Not only is the purity of the final counting sample important, but the materials which make up the scintillator solution must also be of high purity. The purity of the starting material will influence the recovery, and hence the final count measurement.

If labeled compounds are used as starting materials in an experiment, it may be necessary to purify the labeled compound before starting. Many labeled compounds, including ^{3}H labeled, undergo autoradiolysis upon storage. This can lead to a loss of activity, nonspecific labeling, and often side reactions. The method of storage of labeled compounds can often determine the success of an experiment. Storage at room temperature or in a frozen state often leads to high rates of decomposition. The rate of self-radiolysis is reduced by reduction of storage temperature. But, upon freezing of the sample, the rate of decomposition is greatly accelerated due to local high concentration of free radicals and excited molecules which react with the sample. Table VI-1 lists the data on decomposition of a labeled steroid as a function of storage time at three different temperatures (1).

Table VI-1

Percentage decomposition of a ^{3}H-labeled steroid as a function of storage temperature and time

Temperature (°C)	Physical state	% Decomposition after		
		2 months	18 months	20 months
Room temperature	Liquid	2	50	75
0 to +5	Liquid	1	8	15
−5 to 0	Solid	5	90	100

There are many techniques for purifying the sample for liquid scintillation counting. It is beyond the scope of this book to try to discuss all of these methods. Instead, the sample preparation discussions deal with the introduction of various types of samples into liquid scintillator solutions for counting.

Direct Solubilization

Many organic compounds can be directly dissolved in the liquid scintillator solution. These are compounds which are soluble in the solvents used in liquid scintillator solutions, the aromatic hydrocarbons (benzene, toluene, *p*-xylene, and 1,2,4-trimethylbenzene) and dioxane. Often a limiting feature of this method is the sample solubility. Another important factor is the effect of the sample on solute solubility. Early users of *p*-terphenyl as a scintillator solute often noticed the formation of crystals of *p*-terphenyl at the bottom of the counting vial after the addition of certain types of samples. This was due to the fact that to be an efficient scintillator the *p*-terphenyl concentration had to be very near the solubility limit of *p*-terphenyl. Any changes in solvent makeup usually led to precipitation of *p*-terphenyl from the solution.

Certain gases also fall into the category of compounds that are directly soluble in aromatic solvents. Some gases which have been successfully counted directly in liquid scintillator solutions with aromatic solvents are listed in Table VI-2. The basic process by which these gases were counted

Table VI-2

Some radioactive gases counted by direct solubilization in liquid scintillation solutions

Isotope	Compound	Type of radiation	Energy (MeV)	Ref.
^{3}H	H_2(TH)	β^-	$E_{max} = 0.018$	a
^{14}C	$CO_2(^{14}CO_2)$	β^-	$E_{max} = 0.156$	b
^{35}S	$H_2S(H_2\,^{35}S)$	β^-	$E_{max} = 0.167$	c
^{85}Kr	Kr	β^-	$E_{max} = 0.672$	d,e
^{131m}Xe	Xe	e^-	0.134, 0.164	d
^{211}At	At	α	5.89, 7.43	f
^{222}Rn	Rn	α	5.49	d

[a] M. L. Curtis, S. L. Ness, and L. L. Bentz, *Anal. Chem.* **38**, 636 (1966).
[b] D. L. Horrocks, *Int. J. Appl. Radiat. Isotop.* **19**, 859 (1968).
[c] B. E. Gordon, H. R. Lukens, and W. Ten Hove, *Int. J. Appl. Radiat. Isotop.* **12**, 145 (1961).
[d] D. L. Horrocks and M. H. Studier, *Anal. Chem.* **36**, 2077 (1964).
[e] R. E. Shuping, C. R. Phillips, and A. A. Moghissi, *Anal. Chem.* **41**, 2082 (1969).
[f] J. K. Basson, *Anal. Chem.* **28**, 1472 (1956).

with high efficiency was the choice of conditions which favored the solubility of the gas in the solvent (2). The solubility is dependent on the equilibrium between the gas in the vapor phase and that dissolved in the solvent:

$$G_{vap} \underset{k_{-1}}{\overset{k_1}{\rightleftharpoons}} G_{sol}$$

and the distribution coefficient K at equilibrium is

$$K = [G]_{sol}/[G]_{vap}.$$

The concentration of gas in the solvent was increased by making the concentration of the gas in the vapor phase very high. This was accomplished by having a very small volume of space above the solvent. Also since the value of K increases with decreasing temperature, the amount of gas in the solvent was increased by counting in a reduced temperature system. Table VI-3 lists some values of K measured for these gases in toluene (2, 3).

For a scintillation vial which contained a given volume of liquid scintillator solution (toluene solvent) the amount of CO_2 dissolved will be a func-

Table VI-3

Solubilities of some gases in toluene as measured with a liquid scintillation system and
radionuclide tag of gas[a]

	K at			
Gas	−20°C	−15°C	24°C	27°C
CO_2	4.4	—	2.2	—
Kr	—	0.9	—	—
Xe	—	5	—	3
Rn	—	32	—	—

[a] See D. L. Horrocks, *Int. J. Appl. Radiat. Isotop.* **19,** 859 (1968); D. L. Horrocks and M. H. Studier, *Anal. Chem.* **36,** 2077 (1964).

tion of the space above the solution and the temperature. Table VI-4 shows
a calculation of the percentage of CO_2 in the toluene-based scintillator
solution on the values of K from Table VI-3. Most of the gases produced
little or no quenching when dissolved in the scintillator solution. A volume
of 2.0 cm³ (STP) of CO_2 ($^{14}CO_2$) dissolved in 10 ml of toluene-based liquid
scintillator solution gave a 95% counting efficiency in a commercial coincidence-type counter. This compared favorably with the efficiency for ^{14}C-
toluene in the same system.

Table VI-4

Distribution of CO_2 between toluene and free space above toluene[a]

Volume of solution (ml)	Volume of free space (ml)	% of the CO_2 in the toluene at	
		24°C	−20°C
0.25	0.005	99.1	99.5
	0.01	98.2	99.1
	0.05	91.7	95.7
2.0	0.005	99.9	99.9
	0.01	99.8	99.9
	0.05	98.8	99.8
10.0	0.10	99.5	99.8
	0.50	97.8	98.8
	1.00	95.7	97.8

[a] See D. L. Horrocks, *Int. J. Appl. Radiat. Isotop.* **19,** 859 (1968).

In some cases samples as a whole may not be directly soluble in the scintillator solvent, although a fraction of the sample is soluble. Often if the activity to be measured is in the soluble fraction, it can be extracted by washing the sample with the scintillator solvent or even the scintillator solution.

Solubilization Techniques

For those samples that are not directly soluble in the scintillator solution, many can be treated in such ways as to make them soluble. However, one precaution to keep in mind, which cannot be stressed too strongly, is that some chemical reactions may lead to the loss of the radioactive nuclide that is being monitored. The location of the label at certain reactive positions can be undesirable. Consider a labeled organic acid which may undergo decarboxylation:

$$R-C^*OOH \xrightarrow{\Delta} R-H + C^*O_2\uparrow.$$

The following subsections deal with the various solubilization techniques: direct action of a solubilizer, complexing agents, secondary solvents, acid solubilization, synthesis, and isotopic exchange.

Direct Action of Solubilizers

Most macromolecules are not directly soluble in a scintillator solution. However, the action of certain chemicals on the chemical bonds of these macromolecules will cause a degradation into smaller, simpler subunits which can be directly dissolved in the scintillator solution. However, the remaining products often have to be solubilized by the use of secondary solvents (discussed later). Table VI-5 lists a few of the types of compounds that have been used as direct solubilizers.

There are several different quarternary ammonium salts commercially available. Most of these are very excellent solubilizers for such materials as tissue, amino acids, proteins, plasma, serum, nucleic acids, and so on. The usefulness of these is closely determined by their purity. Impure quarternary ammonium salts have associated with the final counting solution large amounts of chemiluminescence and quenching. Some of the commercially available solubilizers of this type are

Hyamine Hydroxide, Hyamine 10-X (Rohn & Haas, Inc.),
NCS (Nuclear-Chicago Corp.),
Soluene (Packard Instrument Co., Inc.),
Protosol (New England Nuclear),
Tissue Solubilizer (Eastman Kodak Co.).

Table VI-5

A few of the types of solubilizers used in liquid scintillation counting sample preparation techniques

Type of solubilizer	Example
Quarternary ammonium salts (hydroxide, chloride)	$\left(\begin{array}{c} CH_3 \\ | \\ R'-N-R'' \\ | \\ CH_3 \end{array} \right)^{+}$ Cl^- or OH^-
	R' and R'' are usually straight-chain aliphatic groups of C_6-C_{20} length
Formamide	$H-\overset{\overset{\displaystyle O}{\|\|}}{C}-NH_2$
Formic acid	$H-\overset{\overset{\displaystyle O}{\|\|}}{C}-OH$
Methanolic–potassium hydroxide	KOH in CH_3OH
Enzymatic	Specific enzymes
Base	NaOH or KOH
Acidic peroxide	$HClO_4-H_2O_2$

There is evidence that many of these solubilizers react with PBD and its derivatives (4–6). The result of the reaction is the production of a yellow color which causes a quenching. The different solubilizers give different degrees of quenching, which suggests that the reaction may involve impurities (6). The yellow color is not evident when PPO or PPO–POPOP is used as the scintillator. For some of the commercial quarternary ammonium salts there is as much as a 50% decrease in counting efficiency for ^3H if butyl-PBD is used as the scintillator solute as compared to PPO.

Complexing

The more common methods employing complexing of the sample usually involve techniques for introducing inorganic cations and anions into the toluene scintillator solutions (7). One technique used to accomplish this is through the formation of a salt of an organic acid. Salts of such acids as octoic, alkyl phosphoric, naphthenic, and a few others have been used successfully, some of which are listed in Table VI-6.

Table VI-6

Some organic acids used to dissolve inorganic cations (as salts of the acids) in
organic scintillator solutions

Acid	Formula	Salts of
2-Ethyl hexonoic (octoic)	H_3C—CH—$(CH_2)_3$—$COOH$ $\mid$ CH_2CH_3	Cd, Pb, Bi, U, Ca, Sm, Na, K, Cs, Rb
Alkyl phosphoric		U, Th, Pu, Am, Cf, Fe, Ru, Rh, Ni, Rare earths, Cd, Sr, Y, Nb, Tc, In, Sn, Zr, Ti
Monoalkyl RO—$P {\displaystyle \stackrel{O}{\underset{OH}{\Big\langle}}}$ OH	$R =$ —C_4H_8 (-butyl) —C_8H_{16} (-octyl) —CH_2CH—C_4H_8 $\mid$ C_2H_5 (2-ethylhexyl)	
Diakyl R—O $\diagdown$ P R—O $\diagup$ $\diagdown$ OH (with O double bond)		
Naphthenic	(structure: cyclohexane ring with CO_2H, H_2C, CH_2, H_2C, CH_2, C, H_2)	Pb, Bi, U
Ethylenediamine tetraacetic acid		Pr, Pm, Sr, Y

Another technique involves the formation of complexes with neutral
organic compounds and dissolving the complexes in the liquid scintillator
solution. Some of these types of complexes are

  (a) Trialkyl phosphates, $(R—O)_3$–P— O,
  (b) Tri-*n*-octylphosphine oxide (TOPO),
  (c) Tri-*n*-caprylamine.

Secondary Solvents

Quite often an insoluble sample can be dissolved in a solvent other than
the scintillator solution solvent, and then, while in this secondary solvent,
it can be dissolved in the scintillator solvent. There are several drawbacks
to this technique:

1. Dilution of the scintillator solvent with large amounts of secondary solvent will decrease the energy transfer efficiency (reduce counting efficiency);

2. Dilution of sample–secondary solvent with the scintillator solution often renders the sample once again insoluble;

3. Dilution of the scintillator solvent with the sample–secondary solvent can cause a decrease in the solubility of the solutes, and in some cases actually cause a crystallization of solutes;

4. A secondary solvent is often a quencher.

Some secondary solvents which have been used with different degrees of success are alcohols (methanol, ethanol, etc.), glycol ethers, and water (8, 9). Alcohols and glycol ethers have been used to solubilize small amounts of water and aqueous solutions into aromatic solvents. With ethanol–aqueous it is necessary to dissolve the aqueous in the ethanol prior to injection into the scintillator solvent. Injection of aqueous into alcohol–toluene mixtures does not lead to complete solubilization with the same amount of alcohol.

Dioxane is a unique solvent in that water and many aqueous solutions are directly soluble in it. Also, the spectral distribution of photons from the dioxane water system shifts to longer wavelength with increasing concentration of water (10). Thus the overlap of dioxane–water emission with solute absorption is increased (i.e., the energy transfer is more efficient) with increasing water concentration. Figure VI-1 shows the shift of the spectrum with water concentration.

The sample (or salt) concentration in the water, which is used as a secondary solvent, can be important in the stability of the scintillation mixture.

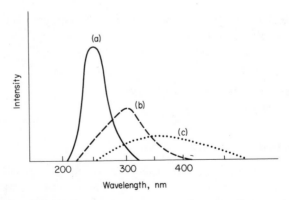

Fig. VI-1. Fluorescence spectra of neat *p*-dioxane and water–*p*-dioxane mixtures. (a) Neat *p*-dioxane, (b) 1% water, (c) 10% water (10).

High salt or sample concentrations may result in precipitation when the water is diluted by the dioxane solvent. Not only will the salt precipitate, but often the sample is carried down also. This leads to a two-phase system with resulting low counting efficiency.

Acid Solubilization

Many samples can be converted to soluble products by the action of acids. Usually oxidizing acids are employed. Under favorable conditions the organic material can even be completely converted into CO_2 ($^{14}CO_2$) and H_2O (THO). Some of the oxidizing acids which have been used with certain types of organic materials are HNO_3, $HClO_4$, mixtures of $HClO_4$–H_2O_2, persulfate, bromine, and mixtures of H_2SO_4–periodate–chromic acid (11–23).

In some cases the acids are used to liberate a part of the organic material, usually as CO_2 (23–27). This technique has been used successfully to liberate CO_2 from many biological fluids, such as blood, urine, etc. The sample is suspended in a small cup over a trap containing an absorbent for the CO_2. Acid is introduced into the sample cup, and the liberated CO_2 is absorbed. The absorbent and CO_2 are then dissolved in a liquid scintillator for counting the $^{14}CO_2$ present. Figure VI-2 is a simplified diagram of one type of apparatus used. In some cases the CO_2 absorbent is present in the liquid scintillator and the CO_2 is directly absorbed.

In one study (23), several different types of tissue were treated with colorless concentrated HNO_3 and directly dissolved in dioxane base scintillation solution. Successfully digested and counted were rat brain, skeletal muscle, blood plasma, packed blood cells, whole blood, fat, stomach,

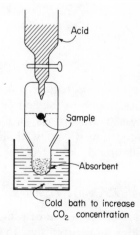

Fig. VI-2. Simplified apparatus for liberation of $^{14}CO_2$ from samples and collection in absorbent (ethanolamine, phenethylamine, or hyamine hydroxide) for liquid scintillation counting.

Acid

Sample

Absorbent

Cold bath to increase CO_2 concentration

small intestine, cecum, large intestine, liver, kidney, eyes, testes, spleen, and skin. Using acid eliminated the luminescence that is normally present with basic digests in dioxane.

Synthesis

Some samples have been converted into CO_2 ($^{14}CO_2$) and H_2O (THO) from which benzene or toluene has been synthesized (28–35). In this manner a nonquenching form of the radioactivity (^{14}C or ^3H) has been prepared. This technique has been used almost exclusively in cases where the amount of radioactive nuclide is very low per weight of sample. The dating of old materials is one application of this technique. The conversion efficiency for $^{14}CO_2$ into benzene or toluene is almost 100% with essentially no isotope effect ($^{14}C:^{12}C$). However, the conversion efficiency for tritiated water into benzene is at most 50%, and there is some isotopic effect.

Of course, many other compounds containing a radioactive nuclide can be synthesized directly. Iodine can be added to toluene for incorporation of radioiodine into liquid scintillators (36). The addition of other halogens across double bonds of unsaturated hydrocarbons can be used to solubilize these radiohalogens in liquid scintillator solutions.

If the hydrogen (tritium) of the sample can be converted to hydrogen gas (H_2 and TH), this can be reacted with some unsaturated hydrocarbons to form saturated hydrocarbons which are directly soluble in the liquid scintillator (37, 38). Conversion of sample hydrogen is often difficult. But most can first be converted to water, and the water can be electrolyzed to H_2 and O_2. Some reactions that have been used are

$$CH\equiv C\!-\!C\equiv CH + 4H_2 \longrightarrow CH_3\!-\!CH_2\!-\!CH_2\!-\!CH_3$$
$$n\text{-Butane}$$

$$\underset{\underset{CH_3}{|}}{CH_2}\!=\!C\!-\!C\equiv CH + 3H_2 \longrightarrow CH_3\!-\!\underset{\underset{CH_3}{|}}{CH}\!-\!CH_2\!-\!CH_3$$
$$\text{iso-Pentane}$$

These reactions require a catalyst that can be a form of palladium, and the yields can be as high as 100%. One disadvantage is the low boiling point of these hydrocarbons—n-butane, $-0.6°C$, and isopentane, $28°C$. Also, incorporation of a few milliliters of these hydrocarbons into a toluene scintillation cocktail reduces the tritium counting efficiency to about 25%.

Isotopic Exchange Techniques

Another technique of sample preparation, which is not often employed but does show promise with certain types of samples, is isotopic exchange

(39–44). To date this technique has been used primarily for the determination of tritium in water. There are problems with isotopic reaction rate factors for exchange of hydrogen 1 and tritium (hydrogen 3). In most cases elevated temperatures and metal catalysts are required to achieve equilibrium in reasonable time.

In the exchange of tritium between water and a hydrocarbon several factors are important, including the choice of the catalyst. If an acid is used as the catalyst, corrections are needed for the amount of hydrogen introduced as acid. The number of exchangeable hydrogen atoms per molecule and the effects of impurities (in the hydrocarbon, the water, and/or catalyst) are important in determining the final concentration of tritium in the hydrocarbon. It is also important to choose a hydrocarbon that will be an efficient scintillator solvent so that the maximum amount of sample may be counted.

Benzene, toluene, o-xylene, and p-xylene have been used for catalytic exchange for two reasons: All their hydrogen atoms are exchangeable and these compounds are good scintillator solvents (45–48). These exchange reactions are metal catalyzed. For equal volumes of water, an aromatic hydrocarbon exchange requires temperatures of 130°C for 2 days. The water has to be pure, as even small amounts of impurities will poison the catalyst (43).

For impure aqueous samples, i.e., urine and plasma, it is possible to utilize the acid-catalyzed exchange with cyclohexene (42, 49, 50). The tritium-containing water is hydrolyzed with such compounds as acetyl chloride, PBr_3, or PBr_5 which produce the halogen acid. The exchange reaction by the addition–elimination reaction is

$$\text{TBr} + \underset{\substack{| \\ \text{H}}}{\overset{\substack{\text{H} \\ |}}{\text{C}}} = \underset{\substack{| \\ \text{H}}}{\overset{\substack{\text{H} \\ |}}{\text{C}}} - \;\rightleftharpoons\; -\underset{\substack{| \\ \text{H}}}{\overset{\substack{\text{H} \\ |}}{\text{C}}} - \underset{\substack{| \\ \text{T}}}{\overset{\substack{\text{Br} \\ |}}{\text{C}}} - \;\rightleftharpoons\; -\underset{\substack{| \\ \text{T}}}{\overset{\substack{\text{H} \\ |}}{\text{C}}} = \underset{\substack{| \\ \text{H}}}{\overset{\substack{\text{H} \\ |}}{\text{C}}} - \; + \text{HBr}.$$

The exchange requires temperatures above 130°C for several days to reach equilibrium. The yield is reduced due to side reactions such as the formation of cyclohexyl bromide.

Since cyclohexene is not an efficient scintillator solvent by itself, it is necessary to dilute with toluene or xylene to obtain a reasonable counting efficiency. This in effect reduces the amount of tritium which can be detected. A 50:50 mixture of cyclohexane–o-xylene has a tritium counting efficiency of 47%, whereas neat cyclohexene has a 14% tritium counting efficiency.

Exchange times of a few hours can be utilized if equilibrium obtainment is not necessary. For nonequilibrium exchanges all reaction conditions have to be carefully controlled, otherwise differences in measured activity may not be due to real differences in the water. Exchange times of only a few hours have been used in some cases.

Heterogeneous Samples

Almost since its first use as a high-efficiency counter of samples dissolved in the liquid scintillator, other groups realized the potential of liquid scintillator systems for the measurement of samples that were not soluble in the solution. Thus today there is probably more use of liquid scintillation counting with nonsoluble samples than there is with soluble samples.

Filter Disks and Paper Strips

In many experiments the final sample can be collected on a filter disk or as a spot on a paper chromatogram (51–72). These disks or spots are often placed directly into the counting vial containing the scintillator solution. Of course, this technique is subject to many uncertainties which are more critical for the low-energy beta emitters. Several factors which will affect the measurements are:

(a) the amount of material on the paper (self-absorption),
(b) the thickness of the paper,
(c) the solubility of the sample or other material present with the sample on the paper, and
(d) the orientation of the paper (light scattering).

Relative counting with filter disks can be used if the counting conditions are reproducible. The position and/or orientation of the paper or disk is critical for both low- and high-energy beta emitters (73–75). For low-energy emitters, if the deposit is turned downward or next to the glass wall, the count rate will be reduced. Also, for high-energy beta particles, close proximity to the vial walls will increase the chance of the beta particle giving all or part of its energy to the wall rather than to the scintillator solution (wall effect). If the filter disk is placed on the bottom of the vial, the maximum counting efficiency that could be obtained should be near 50% for energetic beta emitters, decreasing with decreasing beta energy.

Some filter disks have been employed which dissolve in the scintillator solution (76). These eliminate the problems due to the presence of the disk. However, if the sample is not soluble in the solvent, the sample will merely settle to the bottom of the vial. The use of suspending agents will maintain the sample dispersed in the solvent.

If part of the sample is soluble in the solvent, the total counting efficiency will be due to two different counting efficiencies. The part that dissolved will have one efficiency, while the part still on the disk will have a different efficiency (usually lower). If the sample is completely insoluble, it is possible to use an external standardization technique (external standard channels ratio) to monitor changes in the scintillation cocktail. This is important because small amounts of soluble impurities can cause quenching in the scintillation solution which will decrease the measured count rate of the sample on the disk. Thus, two identical samples on filter disks will give different count rates if a small amount of solvent-soluble quencher is added to one vial.

The thickness and the material makeup of the filter paper or disk will also affect the counting efficiency. For glass filters the efficiency can be markedly different for material on the surface as opposed to that embedded in the pores. This is very critical for the low-energy beta particles from ^3H.

In many cases reproducible counting efficiencies can be obtained by addition of a carrier of known weight (many times more than the sample). This will give the same amount of self-absorption for each sample. Of course, the carrier has to be added before the filtration step, and time must be allowed for complete mixing with the real sample.

The distribution of pulses produced by a sample on a disk or filter paper will be quite different than that produced by a dissolved sample. Beta particles from the surface will be unattenuated; beta particles from within the sample will be partially attenuated; and, depending on the depth at which they originate, some of the low-energy beta particles will be completely absorbed in the sample. This tends to favor the detection of high-energy particles and the elimination of low-energy beta particles. However, the degraded high-energy beta particles will produce pulses like low-energy beta particles. Thus the actual spectrum cannot be predicted but will depend on self-absorption and surface area.

In one study (51), the filter disks were merely dipped in a scintillator solution, completely wetted, and not allowed to dry. The scintillator-impregnated disks were placed directly on the face of the multiplier phototube. Very low ^3H counting efficiencies were obtained, but satisfactory ^{14}C and ^{32}P efficiencies were realized. It was even possible to do some dual label determinations.

The counting of filter paper samples of radon and its daughters was accomplished by merely dropping the disks into a vial containing a scintillator solution (77). However, because of the shorter range of alpha particles compared with beta particles, more of the energy of an alpha particle was expended in the disk than was that of a comparable beta particle (0.1 times the alpha-particle energy). Thus it was difficult to resolve pulses due to alpha particles from those due to beta particles. Utilizing this difference in range was the basis of a method for discriminating between alpha and beta particles. Using a very thin scintillator solution surrounding the disk, the alpha particles were stopped in a short distance, whereas the lower specific ionization beta particles passed through the scintillator without causing interactions.

Other Solid Supports

The counting of materials separated by thin-layer chromatograms (TLC) can be done by cutting out the zone containing the sample and placing the sample and backing material directly into a counting vial containing the scintillation solution (78, 79). It is important to be sure that if eluting solvents and spot developers do dissolve in the scintillator solution, they do not cause severe quenching.

Proteins and nucleic acids can be separated and isolated by electrophoretic techniques using polymerized acrylamide gels. The gel can be sectioned and the activity in the sections measured by placing the gel section in a liquid scintillator solution. Most often the gels are dissolved, but if the slices are very thin, the activity can be measured directly. If the slices are of uniform thickness, the measured count rate will be proportional to the activity in the slice. Again the elution solvents and impurities, if dissolved in the scintillator solution, may cause quenching.

Suspension of Solids

It has long been realized that radioactive-labeled solids which are completely insoluble in a liquid scintillator solution can be counted with reasonable efficiency by merely dispersing the solids, in a very fine powder form, in the scintillator solution (24, 80–87). However, the main disadvantage was a changing count rate, with time, due to the fine powder settling to the bottom of the counting vial.

There were two problems that had to be tackled before suspension counting of solids could be used routinely: (a) prevention of the settling of the solid, and (b) uniform, small size of the solid particles to give a minimum of self-absorption, which was constant.

In many early techniques the solids were actually ground in a mortar. This did not always lead to uniform particle size, and also gave rise to difficulty in recovery of all the radioactive solids from the mortar and pestle. If sufficient sample is available, a small fraction of the powder can be selected by using screens of standard mesh. The particle size is important because as the size increases the surface area decreases, and thus more of the radioactivity will be within the solid. Beta particles originating inside the solid will lose part of their energy before they reach the liquid scintillator solution. Indeed, some of the beta particles will be completely stopped in the solid. The number that will be stopped increases as the size of the particles increases. Those stopped in the particles or reduced in energy below a minimum detectable energy will not be counted. Table VI-7 shows data obtained with labeled d-glucose to show the effect of particle size on the counting efficiency of ^{14}C (88).

Table VI-7

Self-absorption losses of ^{14}C beta particles as function of the physical state of sample of ^{14}C-labeled glucose[a]

Nature of the d-glucose-$^{14}C(U)$ sample	Relative count rate for equal amount of ^{14}C
In homogeneous solution	1.00
Fine powder suspension	0.84
Coarse powder suspension	0.72
Crystals suspension	0.55

[a] See J. C. Turner, "Sample Preparation for Liquid Scintillation Counting." Radiochem. Centre, Amersham, England, 1967.

Originally solid samples were suspended merely by shaking the solid–liquid scintillator solution and following the count rate as a function of time to correct for the settling rate. This technique, while providing adequate results, was subject to too many uncertainties. The size of the particles not only affected the settling rate but also the counting efficiency.

Methods for obtaining a uniform particle size and settling rate led to the use of sonic methods for dispersing solids in the immiscible solutions (89). The sonic dispersion produced very small particles which could be redispersed after settling merely by shaking with reproduction of counting efficiencies obtained immediately after the sonic treatment. In one report (89) up to 40 mg of $BaCO_3$ tagged with ^{14}C was sonically dispersed in a toluene liquid scintillator solution giving 66% counting efficiency with essentially no settling within 30 min after preparation. Several months later a simple hand shaking reproduced the identical results.

To eliminate these settling effects, agents were utilized to increase the viscosity of the scintillator solutions. Two agents used, which did not cause quenching, were aluminum stearate and various forms of finely dispersed silica. Other agents that have been used are "Thixcin" (ricinoleic acid), toluene diisocyanate, and polyolefin resins. Many of these agents actually form gels which support the solid particles in the scintillation medium. Heat is often used to cause the gel formation.

Many different materials have been counted by this solid suspension technique. Most frequently it has been used for ^{14}C determinations after collection of $^{14}CO_2$ as $BaCO_3$ or Na_2CO_3. Other nuclides that have been measured include ^{45}Ca as $CaCO_3$ or CaO, ^{55}Fe or ^{59}Fe as ferriphosphate salts (90), ^{90}Sr as $SrSO_4$, and ^{137}Cs as Cs salts. Inorganic residues from combustion of biological materials have been counted by their suspensions. Also, dried tissue has been counted, provided it does not dissolve or partly dissolve in the scintillator solution.

Emulsions

A common misconception about emulsions is that they are true solutions. The truth is that emulsions are really heterogeneous mixtures. An immiscible solvent is dispersed in another solvent by use of the emulsifier agent. Aqueous solutions can be dispersed in toluene scintillator solutions, or at the other extreme, toluene scintillators can be dispersed in an aqueous solution. The most common types of emulsifiers used in liquid scintillation counting are polyethoxylated surfactants (91). They contain a hydrophilic part (dissolving in water) and a hydrophobic part (dissolving in toluene):

$$C_m H_{2m+1} \underset{\text{Hydrophobic}}{—\bigcirc—} \underset{\text{Hydrophilic}}{(OCH_2—CH_2)_n—OH}.$$

The size of the micelles of water or aqueous solution in the liquid scintillator solution is very important in the determination of counting efficiency. This is more important for low-energy events. Thus the effect of micelle size on tritium is more important than on ^{14}C. The size may vary with the amount of water in the scintillator solution. If the water concentration becomes too great, the micelles may actually combine and cause a phase separation. The appearance of the emulsion may be due to the number and size of the micelles. A greater number and size of micelles will produce a milky appearance due to light scattering.

Many investigations have been reported which deal with the optimum concentration of emulsifier and water for a stable counting solution with a high figure of merit (91–108). The figure of merit is defined as the product of the volume of sample and the counting efficiency:

figure of merit = (volume of sample) × (counting efficiency for radionuclide).

In one such study (103, 104) it was reported that an improved figure of merit was obtained by the use of

   (a) *p*-xylene as the solvent (instead of toluene),
   (b) bis-MSB as the secondary solute,

(c) Triton N-101 surfactant (instead of Triton X-100), and

(d) polyethylene vials.

The figure of merit $E \times M$, where E is the tritium counting efficiency and M the volume of water (in milliliters), was measured as a function of water content of the scintillator solution at different ratios of p-xylene to Triton N-101 (104). The results are shown in Fig. VI-3.

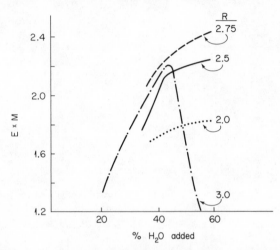

Fig. VI-3. The figure of merit ($E \times M$) as a function of the added water with different amounts of emulsifier (Triton N-101). R = volume of p-xylene/volume of Triton N-101 (104).

In another study (97) the relative count efficiency was measured (cpm/ TU, where TU is tritium units, defined as one tritium atom per 10^{18} ^{1}H atoms) as a function of water content with different amounts of Triton X-100 in toluene. The results are shown in Fig. VI-4.

The Triton X-100–toluene–water system is not a simple system. There seems to be several changes in the physical nature of the solution. These changes in many cases affect the counting properties of the solutions. In one study (100) the scintillation solution consisted of toluene and Triton X-100 in the ratio of 2:1 v/v, respectively, to which varying amounts of water were added. In Fig. VI-5 the counting efficiency and general appearance are given as a function of the amount of water added.

The counting efficiency of an emulsion system was measured using ^{3}H-toluene and THO. A commercially prepared emulsion system (Beckman Ready-Solv VI) was used (109), which is a toluene–PPO–M_2-POPOP solution containing 16% BBS-3 (a special highly purified emulsifier). The water content was varied from 0 to 3 ml in a total volume of 15 ml, and the counting volume was kept constant to eliminate any possible effects

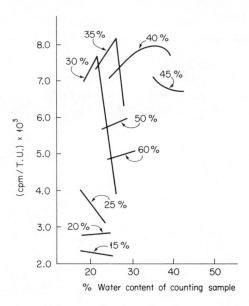

Fig. VI-4. Detection sensitivity as a function of water content, where percentage values indicate Triton X-100 content of the Triton X-100/toluene part of counting solution (97).

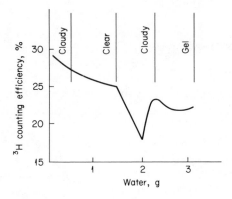

Fig. VI-5. The ³H counting efficiency and general physical appearance of 2:1 toluene–Triton X-100 as a function of water added to 10 ml of scintillator solution (100).

due to light collection differences in the counting chamber. Table VI-8 lists the results obtained. The scintillator quench was monitored by a measure of the Compton edge pulse height for a $^{137}Cs–^{137m}Ba$ source (662-keV γ rays) (110). The results indicate that the water phase and the toluene phase are completely separate. Increased water quenches the counting efficiency of ³H-toluene only slightly. Also, the counting efficiency

Table VI-8

Counting efficiency of ³H-toluene and ³H-water as a function of water content[a]

Water (ml)	Ready-Solv VI (ml)	³H efficiency (%)	Compton Edge RPH
	50 λ of ³H-toluene in Ready-Solv VI		
0	15	42.4	187.5
0.05	15	42.3	187.5
0.5	14.5	42.2	185.5
1.0	14	40.5	184.5
1.5	13.5	39.5	184
2.0	13	39.3	184
2.5	12.5	41.8	186.5
3.0	12	40.5	183.5
	50 λ of THO in Ready-Solv VI		
0	15	—	186
0.05	15	41.7	186
0.5	14.5	39.4	184
1.0	14	38.8	183
1.5	13.5	37.6	183
2.0	13	37.4	182
2.5	12.5	37.6	185
3.0	12	36.7	184
	Background (pure water)		
0	15		185.5
0.05	15		185.5
0.5	14.5		183.5
1.0	14		183
1.5	13.5		182.5
2.0	13		182
2.5	12.5		184.5
3.0	12		183.5
3.5	11.5		183

[a] Quench monitored by Compton edge techniques and compared to identical samples with no tritium.

for THO is decreased only slightly from 0.05 to 3.0 ml of water. This indicates that all the micelles of water are the same size and do not coalesce into larger micelles.

Further studies (109) of the complete separation of the aqueous and organic phases were performed using the pulse height distribution to note any changes in the light yield per unit energy deposited in the scintillator solution. First the ³H pulse distribution was measured (with a multichannel analyzer) for THO and varying amounts of water in Ready-Solv VI. Figure

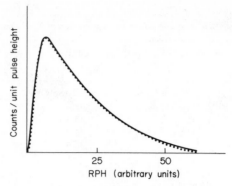

Fig. VI-6. Pulse spectra for tritium beta particles with 50 λ of water and 3.0 ml of water in Ready-Solv VI: − 0.050 ml THO; .... 0.050 ml THO + 3 ml H₂O.

VI-6 shows the results. There is no difference in the pulse height distribution at water contents between 0.05 and 3.0 ml, and surprisingly, the water does not even act as a diluter. There is no interference (quenching) in the scintillation processes. Every concentration of water between 0.05 and 3.0 ml gave the same distribution, within statistical limits.

The lack of quenching with increasing water content was also shown (109) by measuring the pulse height distribution produced by an aqueous solution of NaI containing ^{125}I. The ^{125}I decays by electron capture and subsequent internal conversion. These produce two major essentially monoenergetic groups of electrons, at 12 and 40 keV. Figure VI-7 shows the distribution obtained. The shift of the peaks is very slight, indicating that the scintillator part of the solution is unaffected by the amount of water present. However, if a quencher that is soluble in the toluene phase is present, the scintillation yield is changed the same as if the water were

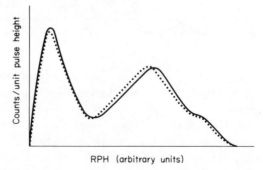

Fig. VI-7. Pulse spectra for ^{125}I electrons with 50 λ and 3.0 ml of aqueous in Ready-Solv: − 50 λ of aqueous Na ^{125}I; ···· 3.0 ml of aqueous Na ^{125}I.

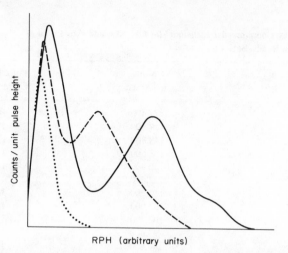

Fig. VI-8. Pulse spectra for ^{125}I electrons as a function of added quencher (CH_3NO_2): — no added CH_3NO_2; ---- 30 λ added CH_3NO_2; ····· 200 λ added CH_3NO_2.

not present (109). Figure VI-8 shows the change in the pulse spectrum for 1.0 ml of aqueous Na ^{125}I with CH_3NO_2 added as a quencher. Besides monitoring the quench effect by the shape of the pulse spectrum, the ^{125}I count rate was obtained (proportional to the area under each spectrum) along with the Compton edge pulse height. These are given in Table VI-9.

From these data it can be seen that the external standard technique (Compton edge) is a very good indicator of the quench level in this system. The data are plotted in Fig. VI-9.

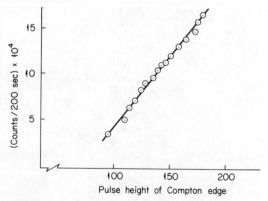

Fig. VI-9. Compton edge quench monitor versus observed count rate for [125]I (1.0 ml aqueous) quenched by added CH_3NO_2 (chemical quench).

Table VI-9

Quench monitor by Compton edge technique[a] for CH_3NO_2 added to 1.0 ml of
aqueous Na ^{125}I in Ready-Solv VI

Added CH_3NO_2 (λ)	Counts in 200 sec	Pulse height of Compton edge
0	173,400	187.5
10	159,500	179.5
20	148,800	171.5
30	138,500	163.5
40	130,400	156.5
50	122,200	150
60	114,300	144
70	107,000	138.5
80	102,200	133.5
90	93,400	129
100	86,850	124.5
110	80,890	121
120	75,390	117.5
130	69,120	114
150	62,350	108.5
200	38,820	94

[a] See D. L. Horrocks, *Nature* (*London*) **202**, 78 (1964).

Table VI-10

Quench monitor by Compton edge technique[a] for methyl orange added to 1.0 ml of
aqueous Na ^{125}I in Ready-Solv VI

Added methyl orange (λ)	Counts in 200 sec	Pulse height of Compton edge
0	174,350	187
10	154,000	176.5
20	137,040	166
30	124,350	157
40	113,160	149.5
50	103,660	143
60	93,150	136.5
70	84,600	131
80	76,520	126.5
90	67,840	121
100	60,150	116
120	48,040	108.5
150	33,920	99.5
200	19,670	90
250	12,750	87

[a] See D. L. Horrocks, *Nature* (*London*) **202**, 78 (1964).

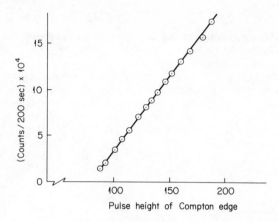

Fig. VI-10. Compton edge quench monitor versus observed count rate for ^{125}I (1.0 ml aqueous) quenched with methyl orange (color quench).

The same experiment was performed using a color quenching agent, methyl orange. The data obtained are given in Table VI-10 and Fig. VI-10. The data indicate that the external standard method, Compton edge pulse height, gave a response almost the same as the chemical quencher, CH_3NO_2. However, on very close examination, the slopes of the two plots are different, indicating a slightly different quench rate. This may be due in part to the addition of higher atomic number atoms (Na) in the methyl orange, which could scatter the γ rays differently than the C, H, O, N atoms of the chemical quencher, CH_3NO_2. The equations for the two quench curves are

Chemical quench: $\text{counts}/10^4 = 0.141$ (Compton edge value) $- 9.1$
Color quench: $\text{counts}/10^4 = 0.158$ (Compton edge value) $- 12.4$.

The size of the micelles of aqueous solutions was also shown (109) to be very small by measuring the alpha particles from ^{241}Am. The range of the 5.5-MeV alpha particles is approximately 3×10^{-3} cm in water. But even more important is the definite range of the alpha particles; i.e., the alpha particles lose almost all their energy at the very end of their track. Any increase in micelle size would alter the detection and pulse height resolution. Figure VI-11 and Table VI-11 show the results for aqueous solutions of ^{241}Am as a function of the amount of water in an emulsifier system (Beckman Ready-Solv VI).

Even though the peak height is greater for the sample with 3.0 ml of water, shown in Fig. VI-11, the distribution is narrower and the total counts are the same in the two distributions. The total counts, as given in Table VI-11, are representative of the area under each peak.

Table VI-11

Counting data from the pulse spectra of ²⁴¹Am alpha particles as a function of water content in Ready-Solv VI

Water (ml)	Scintillator solution (ml)	Counts/100 sec	Pulse height of peak	Width of peak at half-height	Line width (%)
0.05	15	209,077	104	14	13.5
0.5	14.5	210,429	102.5	13.5	13.2
1.0	14	208,000	101.5	13.5	13.3
1.5	13.5	208,230	100.5	13.0	12.9
2.0	13	210,207	100	12.0	12.0
2.5	12.5	211,000	101	12.5	12.4
3.0	12	209,050	99.5	12.0	12.1

Many emulsion systems will change their physical appearance with temperature change. There seems to be a temperature range of stability of the emulsion system. This temperature range was a complex function of the ratio of toluene to emulsifier, the amount of water, and the general method of preparation. Sometimes stability could be induced by heating the mixture until a clear solution was obtained, followed by rapid cooling to form a gel which prevented phase separation. Addition of the aqueous

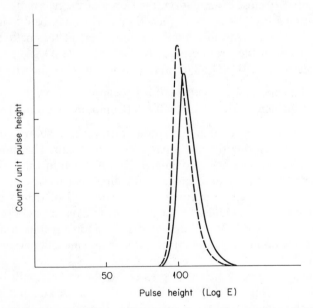

Fig. VI-11. Pulse spectra of ²⁴¹Am alpha particles for samples containing 50 λ (—) and 3.0 ml (----) of water in Ready-Solv VI.

phase to the scintillator solution, as opposed to the opposite order of mixing, seemed to produce more reproducible results.

The range of temperature for a stable emulsion as measured by one investigator (104) is shown in Table VI-12. At these very high water contents (48–32%) there was usually only a 4–8°C variability in the temperature.

Table VI-12

Temperature range of stability of mixture of emulsifier scintillator solution and water[a]

Emulsifier	Solvent	Ratio of solvent to emulsifier	% Water	Temperature range (°C)
Triton N-101	*p*-Xylene	2.75	40	18–26
Triton N-101	*p*-Xylene	2.75	44	20–26
Triton N-101	*p*-Xylene	2.75	48	22–23
Triton X-100	Toluene	2.75	32	19–23

[a] R. Lieberman and A. A. Moghissi, *Int. J. Appl. Radiat. Isotop.* **21**, 319 (1970).

The salt concentration of the aqueous sample is also important in the stability of the emulsion. Not all salts have the same effect. Dilute acids (1–2 N) can usually be incorporated into an emulsion without causing quenching. High acid concentrations, depending on the particular acid, can cause severe quenching. For instance, HNO_3 will produce severe quenching, whereas HCl will not.

The concentration of salts can affect the counting efficiency of some emulsifier systems (105). Table VI-13 shows the effect of the amount of different concentrations of NaCl on the counting efficiency of ^3H in the aqueous sample. Also listed are the effects of some other materials.

Table VI-13

Effect of salt and acid concentration in aqueous part of sample upon tritium counting efficiency

Sample	Tritium counting efficiency (%) at aqueous content of		
	5%	25%	50%
^3H-toluene	42	—	—
0.5 N NaCl	34	27	21
1 N NaCl	34	27	22
0.5 N NaOH	39	26	21
1.0 N NaOH	29	24	11
40% Sucrose–water	26	25	17
6 N CsCl	36	—	—

Of course the presence of colored material in the aqueous phase will also affect the counting efficiency. Two biological fluids which present a color problem are blood and urine, precluding any decolorization treatment.

It should also be kept in mind that the addition of large amounts of high atomic number atoms to the solution, e.g., Cs, Na, Cl, can cause a greater amount of Compton scattering and even some photoelectric effects for the γ rays from an external standard gamma source. This could give abnormal external standard channels ratio values.

Suspension of Scintillator

In some cases the scintillator material has been suspended in the sample. This often leads to low counting efficiencies, especially for low-energy beta emitters, because the ionizing particles have to travel some distance in a nonscintillation-producing medium before exciting the scintillator. Only those particles that reach the scintillator have a chance of being counted, and only those that reach the scintillator with sufficient energy will actually be counted.

An example of this technique is the use of anthracene crystals dispersed in a flow cell tube (111–117). The sample flows around the packed anthracene crystals giving the effect of a scintillator dispersed in the sample. Counting efficiencies are nominally about 55% for ^{14}C and only 2% for ^{3}H.

Some emulsion systems, when the water content exceeds the scintillator content, can be considered as a suspension of scintillator in the sample (104). Systems with 50% aqueous phase have been shown to have ^{3}H counting efficiencies of 20%.

Higher-energy beta emitters can be counted with efficiencies greater than 50% merely by addition of anthracene crystals to a solution containing the radionuclide (117, 118). The size and suspension of the crystals are important in the efficiency obtained, as is the avoidance of solvents that will dissolve the anthracene crystals or react with the surface layer to cause oxidation or yellowing. Also it is sometimes necessary to add small amounts of detergent to allow the wetting of the anthracene crystals. The counting efficiencies obtained for some radionuclides are listed in Table VI-14.

Similar techniques have utilized beads or fibers of plastic scintillators (119–124) which can be made in almost any size and shape. The plastics, while having a lower scintillation yield, can be prepared with wave shifters, thus reducing the possibility of self-absorption of the emitted photons. Anthracene has a very high self-absorption probability.

Reuse of the anthracene crystals or plastic beads and fibers is dependent on their retention of radioactivity. It has been shown that certain types of

Table VI-14

Some typical counting efficiencies obtained with suspended anthracene craystals[a]

Radionuclide	E_{max}(MeV)	Counting efficiency (%)
^{3}H	0.018	1
^{14}C	0.156	20
^{45}Ca	0.254	49
^{131}I	0.608	58
^{32}P	1.71	78

[a] See E. Rapkin, Lab. Scintillator, Vol. 11, No. 5L, March 1967. Picker Nucl. White Plains, New York, 1967.

material will either be absorbed on the surfaces or in some cases actually react with the scintillator. Counting between use, with only water, will be required to check on retained or trapped radioactivity.

References

1. L E. Geller and N. Silberman, *in* "The Current Status of Liquid Scintillation Counting" (E. D. Bransome, Jr., ed), p. 137. Grune & Stratton, New York, 1970.
2. D. L. Horrocks and M. H. Studier, *Anal. Chem.* **36**, 2077 (1964).
3. D. L. Horrocks, *Int. J. Appl. Radiat. Isotop.* **19**, 859 (1968).
4. A. Dunn, *Int. J. Appl. Radiat. Isotop.* **22**, 212 (1971).
5. A. F. McEvoy, S. R. Dyson, and W. G. Harris, *Int. J. Appl. Radiat. Isotop.* **23**, 338 (1972).
6. D. L. Horrocks, unpublished results, 1972.
7. D. L. Horrocks, Packard Tech. Bull. No. 2. Packard Instrum. Co., Inc., Downers Grove, Illinois, 1961.
8. F. N. Hayes, *Int. J. Appl. Radiat. Isotop.* **1**, 46 (1956).
9. J. D. Davidson and P. Feigelson, *Int. J. Appl. Radiat. Isotop.* **2**, 1 (1957).
10. F. Hirayma, C. W. Lawson, and S. Lipsky, *J. Chem. Phys.* **74**, 2411 (1970).
11. A. Lindenbaum and C. J. Lund, *Radiat. Res.* **37**, 131 (1969).
12. A. Lindenbaum and M. A. Smyth, *in* "Organic Scintillators and Liquid Scintillation Counting" (D. L. Horrocks and C. T. Peng, eds.), p. 951. Academic Press, New York, 1971.
13. B. F. Cameron, *Anal. Biochem.* **11**, 164 (1965).
14. D. T. Mahin and R. T. Lofberg, *in* "The Current Status of Liquid Scintillation Counting" (E. D. Bransome, Jr., ed.), p. 212. Grune & Stratton, New York, 1970.
15. D. D. VanSlyke and J. J. Folch, *Biol. Chem.* **136**, 509 (1940).
16. J. M. Passmann, N. S. Radin, and J. A. D. Cooper, *Anal. Chem.* **28**, 484 (1956).
17. H. Jeffay and J. Alvarez, *Anal. Chem.* **33**, 612 (1961).
18. L. A. Walker and R. Lougheed, *Int. J. Appl. Radiat. Isotop.* **13**, 95 (1962).
19. E. H. Belcher, *Phys. Med. Biol.* **5**, 49 (1960).
20. W. J. Kirsten and M. E. Carlsson, *Microchem. J.* **4**, 3 (1960).
21. B. Bloom, *Anal. Biochem.* **3**, 85 (1962).

22. H. Jeffay, Packard Tech. Bull., No. 10. Packard Instrum. Co., Inc., Downers Grove, Illinois, 1962.
23. M. Pfeffer, S. Weinstein, J. Gaylord, and L. Indindoli, *Anal. Biochem.* **39,** 46 (1971).
24. D. G. Nathen, J. G. Davidson, J. G. Waggoner, and N. J. Berlin, *J. Lab. Clin. Med.* **52,** 915 (1958).
25. D. Cuppy and L. Crevasses, *Anal. Biochem.* **5,** 462 (1963).
26. E. M. Tarjan, *Endocrinology* **88,** 833 (1971).
27. M. J. Woliner, L. Bernal, N. Brachfeld, and E. M. Tarjan, *J. Lab. Clin. Med.* **82,** 166 (1973).
28. M. Tamers, *in* "Organic Scintillators" (D. L. Horrocks, ed.), p. 261. Gordon & Breach, New York, 1968.
29. M. Tamers, *Acta Cient. Venez.* **16,** 156 (1965).
30. J. E. Noakes, S. M. Kim, and J. J. Stipp, *"Proc. Int. Conf. Radiocarbon and Tritium Dating, 6th, 1965,* CONF-650652, p. 68 (1965).
31. J. J. Stipp, E. M. Davis, J. E. Noakes, and T. E. Hoover, *Radiocarbon* **4,** 43 (1962).
32. S. M. Kim, *in* "Organic Scintillators and Liquid Scintillation Counting" (D. L. Horrocks and C. T. Peng, eds.), p. 965. Academic Press, New York, 1971.
33. M. Tamers, Packard Tech. Bull. No. 12. Packard Instrum. Co., Inc., Downers Grove, Illinois, 1964.
34. M. Tamers, R. Bibron, and G. Delibrias, "Tritium in the Physical and Biological Sciences," p. 303. IAEA, Vienna, 1962.
35. M. Tamers and R. Bibron, *Nucleonics* **21,** No. 6, 90 (1963).
36. R. E. Yerick and H. H. Ross, Rep. TID-7688. U. S. At. Energy Commission, Washington, D.C., 1963.
37. H. Perschke, M. B. A. Crespi, and G. B. Cook, *Int. J. Appl. Radiat. Isotop.* **20,** 813 (1969).
38. H. Perschke and T. Flowkowski, *Int. J. Appl. Radiat. Isotop.* **21,** 747 (1970).
39. J. E. Bradley and D. J. Bush, *Int. J. Appl. Radiat. Isotop.* **1,** 233 (1956).
40. P. Avinur and A. Nir, *Bull. Res. Counc. Israel Sect. A* **7,** 74 (1958).
41. I. Dostrovsky, P. Avinur, and A. Nir, *in* "Liquid Scintillation Counting" (C. G. Bell and F. N. Hayes, eds.), p. 283. Pergamon, Oxford, 1958.
42. M. Anbar, P. Neta, and A. Heller, *Int. J. Appl. Radiat. Isotop.* **13,** 310 (1962).
43. G. E. Calf, *Int. J. Appl. Radiat. Isotop.* **20,** 177 (1969).
44. G. E. Calf, *in* "Organic Scintillators and Liquid Scintillation Counting" (D. L. Horrocks and C. T. Peng, eds.), p. 719. Academic Press, New York, 1971.
45. J. L. Garnett, *Nucleonics* **20,** No. 12, 86 (1962).
46. J. L. Garnett and W. A. Sollich, *Aust. J. Chem.* **18,** 1003 (1965).
47. J. L. Garnett and W. A. Sollich, *J. Catal.* **2,** 350 (1963).
48. J. L. Garnett and W. A. Sollich-Baumgartner, *Advan. Catal. Relat. Sub.* **16,** 19 (1966).
49. G. E. Calf, B. D. Fisher, and J. L. Garnett, *Proc. Int. Conf. Methods of Preparing and Storing Labelled Compounds, 2nd, 1966,* p. 689. EURATOM, Brussels, 1966.
50. G. E. Calf, B. D. Fisher, and J. L. Garnett, *Aust. J. Chem.* **21,** 947 (1968).
51. J. Roucayrol, E. Overhouser, and R. Schussler, *Nucleonics* **15,** No. 11, 104 (1957).
52. C. H. Wang and D. E. Jones, *Biochem. Biophys. Res. Commun.* **1,** 203 (1959).
53. W. F. Bosquet and J. F. Christian, *Anal. Chem.* **32,** 722 (1960).
54. E. Rapkin, Packard Tech. Bull. No. 2. Packard Instrum. Co., Inc., Downers Grove, Illinois, 1960.
55. S. Segal and A. Blair, *Anal. Biochem.* **3,** 221 (1962).
56. J. Chrirboga, *Anal. Chem.* **34,** 1843 (1962).
57. H. Takahashi, T. Hattori, and B. Maruo, *Anal. Biochem.* **2,** 447 (1961).

58. C. F. Baxter and I. Senoner, *Anal. Biochem.* **7,** 55 (1964).
59. F. Snyder, *Anal. Biochem.* **9,** 183 (1964).
60. J. W. Davies and E. C. Cocking, *Biochim. Biophys. Acta* **115,** 511 (1966).
61. F. Hutchinson, *Int. J. Appl. Radiat. Isotop.* **18,** 136 (1967).
62. M. Naksbandi, *Int. J. Appl. Radiat. Isotop.* **16,** 157 (1965).
63. M. N. Cayen and P. A. Anastassiadis, *Anal. Biochem.* **15,** 84 (1966).
64. N. B. Furlong, N. L. Williams, and D. P. Willis, *Biochim. Biophys. Acta* **103,** 341 (1965).
65. D. R. Johnson and J. Ward, *Anal. Chem.* **35,** 1991 (1963).
66. D. M. Gill, *Nature (London)* **202,** 626 (1964).
67. C. Slot, *Scand. J. Clin. Lab. Invest.* **17,** 182 (1965).
68. R. A. Malt and W. L. Miller, *Anal. Biochem.* **18,** 388 (1967).
69. D. M. Gill, *Int. J. Appl. Radiat. Isotop.* **18,** 393 (1967).
70. N. B. Furlong, *in* "The Current Status of Liquid Scintillation Counting" (E. D. Bransome, Jr., ed.), p. 201. Grune & Stratton, New York, 1970.
71. E. D. Bransome, Jr. and M. F. Grower, *in* "Organic Scintillators and Liquid Scintillation Counting" (D. L. Horrocks and C. T. Peng, eds.), p. 683. Academic Press, New York, 1971.
72. A. Chakravarti and J. W. Thanassi, *Anal. Biochem.* **39,** 484 (1971).
73. J. W. Geiger and L. B. Wright, *Biochem. Biophys. Res. Commun.* **2,** 282 (1962).
74. R. B. Loftfield, Atomlight, No. 13. New England Nucl. Corp., Boston, Massachusetts, 1960.
75. J. D. Davidson, *Proc. Conf. Organic Scintillation Detectors Univ. of New Mexico, 1960* (G. H. Daub, F. N. Hayes, and E. Sullivan, eds.), TID-7612, p. 232. U. S. At. Energy Commission, Washington, D. C., 1961.
76. A. E. Shamoo, *Anal. Biochem.* **39,** 311 (1971).
77. R. Rolle, *Amer. Ind. Hyg. Ass. J.* **31,** 718 (1970).
78. F. Snyder, *in* "The Current Status of Liquid Scintillation Counting" (E. D. Bransome, Jr., ed.), p. 248. Grune & Stratton, New York, 1970.
79. M. F. Grower and E. D. Bransome, Jr., *in* "The Current Status of Liquid Scintillation Counting" (E. D. Bransome, Jr., ed.), p. 263. Grune & Stratton, New York, 1970.
80. F. N. Hayes, B. S. Rogers, and W. H. Longham, *Nucleonics* **14,** No. 3, 48 (1956).
81. B. L. Funt, *Nucleonics* **14,** No. 8, 83 (1956).
82. C. G. White and S. Helf, *Nucleonics* **14,** No. 10, 46 (1956).
83. D. G. Ott, C. R. Richmond, T. T. Trujillo, and H. Foreman, *Nucleonics* **17,** No. 9, 106 (1959).
84. S. Helf and C. G. White, *Anal. Chem.* **29,** 13 (1957).
85. S. Helf, C. G. White, and R. N. Schelley, *Anal. Chem.* **32,** 238 (1960).
86. H. J. Cluley, *Analyst (London)* **87,** 170 (1962).
87. J. N. Bollinger, W. A. Mallow, J. W. Register, and D. E. Johnson, *Anal. Chem.* **39,** 1508 (1967).
88. J. C. Turner, "Sample Preparation for Liquid Scintillation Counting." Radiochem. Centre, Amersham, England, 1967.
89. J. B. Allred, *Anal. Chem.* **39,** 547 (1967).
90. J. D. Eakins and D. A. Brown, *Int. J. Appl. Radiat. Isotop.* **17,** 391 (1966).
91. S. B. Lupica, *Int. J. Appl. Radiat. Isotop.* **21,** 487 (1970).
92. R. C. Meads and R. A. Stiglitz, *Int. J. Appl. Radiat. Isotop.* **13,** 11 (1962).
93. M. S. Patterson and R. C. Greene, *Anal. Chem.* **37,** 854 (1965).
94. R. H. Benson, *Anal. Chem.* **38,** 1353 (1966).

95. J. D. Van der Laarse, *Int. J. Appl. Radiat. Isotop.* **18**, 485 (1967).
96. P. H. Williams and T. Florkowski, "Radioactive Dating and Methods of Low Level Counting," p. 703. IAEA, Vienna, 1967.
97. P. H. Williams, *Int. J. Appl. Radiat. Isotop.* **19**, 337 (1968).
98. R. C. Greene, M. S. Patterson, and A. H. Estes, *Anal. Chem.* **40**, 2035 (1968).
99. J. C. Turner, *Int. J. Appl. Radiat. Isotop.* **19**, 557 (1968).
100. J. C. Turner, *Int. J. Appl. Radiat. Isotop.* **20**, 499 (1969).
101. P. A. Lindsay and N. B. Kurnick, *Int. J. Appl. Radiat. Isotop.* **20**, 97 (1969).
102. A. E. Whyman, *Int. J. Appl. Radiat. Isorop.* **21**, 81 (1970).
103. A. A. Moghissi, H. L. Kelley, J. E. Regnier, and N. W. Carter, *Int. J. Appl. Radiat. Isotop.* **20**, 145 (1969).
104. R. Lieberman and A. A. Moghissi, *Int. J. Appl. Radiat. Isotop.* **21**, 319 (1970).
105. R. C. Greene, in "The Current Status of Liquid Scintillation Counting" (E. D. Bransome, Jr., ed.), p. 189. Grune & Stratton, New York, 1970.
106. R. J. Obremski, Biomed. Procedures BP-NUC-1. Beckman Instrum., Inc., Fullerton, California, 1972.
107. R. J. Obremski, Biomed. Procedures BP-NUC-2. Beckman Instrum. Inc., Fullerton, California, 1972.
108. R. J. Obremski, Biomed. Procedures, BP-NUC-3. Beckman Instrum., Inc., Fullerton, California, 1973.
109. D. L. Horrocks, unpublished results, 1972.
110. D. L. Horrocks, *Nature (London)* **202**, 78 (1964).
111. D. Steinberg, *Nature (London)* **182**, 740 (1958).
112. D. Steinberg, *Nature (London)* **183**, 1253 (1959).
113. D. Steinberg, *Anal. Biochem.* **1**, 23 (1960).
114. E. Rapkin and L. E. Packard, *Proc. Conf. Organic Scintillation Detectors, Univ of New Mexico, 1960* (G. H. Daub, F. N. Hayes, and E. Sullivan, eds.), TID-7612, p. 216. U. S. At. Energy Commission, Washington, D. C., 1961.
115. E. Rapkin and J. Gibbs, *Nature (London)* **194**, 34 (1962).
116. E. Shram and R. Lombaert, *Anal. Biochem.* **3**, 68 (1962).
117. E. Schram, in "The Current Status of Liquid Scintillation Counting" (E. D. Bransome, Jr., ed.), p. 95. Grune & Stratton, New York, 1970.
118. K. A. Piez, *Anal. Biochem.* **4**, 444 (1962).
119. E. Rapkin, Labo. Scintillator, Vol. 11, No. 5L, March 1967. Picker Nucl., White Plains, New York, 1967.
120. E. Schram and R. Lombaert, *Anal. Chim. Acta* **17**, 417 (1957).
121. E. Schram and R. Lombaert, *Biochem. J.* **66**, 21 (1957).
122. R. Tkachuk, *Can. J. Chem.* **40**, 2348 (1962).
123. A. Karmen, *Anal. Chem.* **35**, 336 (1963).
124. A. K. Burt and J. A. B. Gibson, *U. K. At. Energy Auth. Res. Group Rep.* **AERE-R-4638** (1964).

CHAPTER VII

OXIDATION TECHNIQUES

The insolubility, or quenching, of some types of samples in liquid scintillation systems has led to a continuing search for better techniques for sample preparation. One such procedure is the oxidation of the organic compounds that produce CO_2 ($^{14}CO_2$) and H_2O (^3HHO). If the combustion is complete and the recoveries are quantitative, the combustion method will lead to the very accurate measure of ^{14}C and ^3H in samples which otherwise would be very difficult to assay.

In general, oxidation techniques are considered for different reasons, among which are:

1. if the sample is insoluble or sparingly soluble in the liquid scintillator system,

2. if severe quenching results from addition of the sample to the liquid scintillation system,

3. if large amounts of chemiluminescence or phosphorescence occur as the result of the sample being added to the liquid scintillation system,

4. if a wide variety of sample types are to be compared in an experiment, and

5. if it is necessary to separate the ^{3}H and ^{14}C activities before counting.

Wet Oxidation

Wet oxidation is usually performed with a strong oxidizing chemical in a solution containing the organic material. The use of HNO_3-HClO_4 mixtures will oxidize most organic material to CO_2 and H_2O. This procedure will lead to recovery of the ^{14}C as $^{14}CO_2$, but the tritium will be greatly diluted by the H_2O from the acids. Other strong wet oxidizing systems are listed below. The oxidation reagents have to be strong enough to ensure complete oxidation of the organic material. Incomplete oxidation will lead to less than quantitative recovery.

HNO_3	Fuming H_2SO_4–periodate–chromic acid
HNO_3-HClO_4	Bromine
$HClO_4-H_2O_2$	Persulfate
Dichromates–H_2SO_4	Certain enzymes

Dry Oxidation

Dry oxidation involves combustion of the sample in an atmosphere of oxygen. Usually a catalytic surface is necessary to ensure complete oxidation of the organic material to the highest oxides: CO_2, water, oxides of nitrogen, SO_3, etc. Since the combustion does not involve dilution of the sample, both ^{14}C and ^{3}H can be assayed by the combustion process.

Static Combustion

The bomb and flask combustion techniques are static systems (1–21). The organic material and oxygen are placed in a closed system, and the combustion is started by some external stimulus. The efficiency of static systems depends on the mixing properties. With the static systems it is difficult to separate the CO_2 and H_2O. The combustion bombs usually provide complete oxidation, but one is faced with the problem of introducing the absorbents for CO_2 into the bomb.

The oxygen flask combustion technique had been developed as an extension of the early use of oxygen combustion for measurement of the carbon and hydrogen content of organic samples. The basic concept of the oxygen flask technique is to place the organic material in a flask (or some other sealable container) in an atmosphere of oxygen. The sample is ignited, and the combustion products are trapped in the flask. The absorbents for CO_2 and water are added directly to the flask. Once the products are trapped the liquid scintillator solution is added and the mixture is poured into a vial for counting. If the amount of product is too large, an aliquot is placed in the vial along with the appropriate volume of liquid scintillator solution and the sample counted.

Figure VII-1 shows a typical oxygen flask apparatus. It is essentially a heavy-walled Erlenmeyer flask with side arm. The flask is filled with oxygen by flowing a stream of oxygen gas into the bottom of the flask for several seconds (depending on the volume and flow rate). The sample is placed in a sample holder, one type of which is illustrated in Fig. VII-2. Only solid samples or samples on solid supports can be placed in this type of holder. Other types of holders have been devised for all kinds of samples: absorbed on filter papers, plastic bags, quartz boats, etc. The sample in its holder is introduced into the flask and sealed. In the apparatus shown in Fig. VII-1 the seal is maintained by pressure-sealing a rubber stopper in the opening of the flask. The side arm is closed with a rubber tubing pinched shut with clamp. (This tubing will be used later for introduction of the absorbents.)

The combustion is started by some external stimulus. Some types of initiators are electrical currents, spark sources, and induction heaters, but the most commonly used are infrared light beams. With most samples, once the combustion is started, the oxygen atmosphere will sustain the oxidation reaction. The use of electrical currents and spark sources requires a more elaborate method of bringing electrical leads into the flask, which is

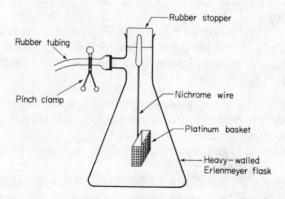

Fig. VII-1. A typical oxygen flask apparatus for the oxidation of organic samples.

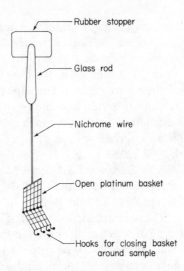

Fig. VII-2.   One type of sample-holding device for oxygen flask apparatus. Platinum basket acts as both sample holder and catalyst for oxidation process.

insulated to prevent shock to the operator, but which is leak-tight to prevent loss of the oxidation products.

After complete combustion of the sample, the gases, CO_2 and H_2O, are cooled and the proper absorbent is introduced into the flask. If water is the desired product, cooling the flask walls to condense the water vapor will allow direct opening of the flask for addition of the water solubilizer. If CO_2 is the desired product, the absorbent must be introduced without loss of CO_2. Several techniques have been reported in the literature.

Since the combustion reaction consumes more oxygen than it produces other gases (mainly CO_2), there is slightly less than atmospheric pressure in the flask after the combustion has cooled. Thus a predetermined amount of additive can be siphoned through the side arm (Fig. VII-1) with very little likelihood of escape of gases from the flask. Other techniques for addition of the absorber are:

(a) by injection by hypodermic syringe through a section cover,

(b) by placement of the absorbent in a breakable container in the bottom of the flask, with breakage after combustion,

(c) by placement of the absorbent in the side arm, and introducing it by tipping or rotating the side arm,

(d) by freezing all gases (dry ice or liquid N_2), opening and adding absorbent, replacing the cap, and allowing to warm to room temperature,

(e) by using a vacuum line connection to pump the CO_2 to a trap containing the absorbent. The water will remain behind if the flask is cooled.

Finally the isolated gases (CO_2 and H_2O), which have been fixed, can be added directly to a vial with appropriate amounts of scintillator solution for counting. If the sample is too large, aliquots of the products can be used. The use of aliquots requires accurate measurement of the sample volume or weight.

A unique modification of the flask technique was developed by Kisieleski (22). Some small samples can be placed in a counting vial which has been filled with oxygen. The cap is secured and combustion started with an infrared lamp. After oxidation is complete, the vial is partly submerged in a dry ice bath (only THO retained) or in liquid nitrogen (both THO and $^{14}CO_2$ retained). After the combustion products are frozen, the cap is removed (allowing the oxygen to escape, which reduces the quench due to excess oxygen) and the liquid scintillation cocktail added. The cap is replaced and the mixture warmed to room temperature. The CO_2 is complexed by some agent, such as ethanolamine, to prevent its escape. Use of certain cocktails will enhance the solubility of water in the organic system.

Gupta (23, 24) has developed a unique modification of the combustion flask technique, called "mylar bag combustion." The sample is placed in an oxygen-filled mylar bag, and the bag is sealed by heat (using a heat sealer like an old-fashioned hair curler). The sample is ignited by using an infrared lamp or (if the mass is sufficient) a microwave generator. After the combustion is completed, the scintillator solution and absorbents are injected through the wall of the bag with a hypodermic syringe. The whole bag is then dropped in a bottle for counting. This method gave a mean tritium counting efficiency of 28% and a mean ^{14}C counting efficiency of 85%. Also it was reported that the background was only 7–13 cpm.

There are several major drawbacks to these static oxidation methods, some of which are listed below:

1. It is often difficult to obtain complete sample oxidation, due to incomplete mixing.

2. Only the sample can be maintained at the high temperature required for combustion.

3. Some volatile gases are distilled from the sample before complete oxidation and condense on the cold walls of the flask.

4. Noncombustiles are still in the flask along with the CO_2 and H_2O, i.e., ^{45}Ca, $^{55,59}Fe$, etc.

5. Incomplete oxidation leads to quenching, and often variable quench between identical samples.

6. It is often necessary to use involved procedures to ensure complete absorption of CO_2, for example, rotation and/or shaking of the flask for long periods.

7. Because of the large volume of oxidation products, it is often necessary to use only an aliquot in order to have a homogeneous counting solution.

Dynamic Combustion

In recent years several very good commercial sample oxidation systems have become available (19, 25–29). These systems are based on a dynamic combustion process in which the oxygen is continually flowing through the combustion chamber, thereby supplying a fresh combustion atmosphere, and at the same time sweeping away the combustion products, i.e., CO_2 and H_2O vapor. The CO_2 and H_2O vapors can then be collected either together or, more important, separately. Thus, starting with a dual-labeled sample, the ^{14}C and ^{3}H can be isolated separately, thus eliminating the uncertainties involved in counting ^{14}C and ^{3}H together.

The sample is placed in a combustion chamber filled with oxygen. The whole chamber is heated, thus preventing condensation of incompletely oxidized products on the walls of the combustion chamber. This leads to more complete oxidation of the sample. Also, some of the oxidizers are supplied with catalysts at the outlet of the combustion chamber to provide complete oxidation of all the gases that are swept out of the combustion chamber.

Besides more complete combustion, the dynamic systems can be adapted to provide complete separation of the CO_2 and H_2O fractions. First, the water can be trapped by condensation and/or freezing of the water. The CO_2 is swept through to a second trap, which is filled with an absorbent which will chemically combine with the CO_2. Such absorbents are usually amines, i.e., ethanolamine or phenethylamine. These amines react with the CO_2 to form a carbonate which is not volatile. Other absorbers of CO_2 are NaOH, $BaSO_4$, Hyamine and Hyamine 10-X, Primene 81-R, etc. These agents are listed in Table VII-1. Of those listed, phenethylamine and ethanolamine are the best, at this time. They give quantitative absorption (as long as excess amine is present), low quench, and a minimum of difficulty in dissolving in toluene liquid scintillator solutions.

It is, of course, very important to ensure that conditions are satisfied, so that all the oxidation products can be collected. Since the water vapor is collected by condensation, there is no limit to the amount that can be

Table VII-1

Some trapping agents for CO_2

Name	Formula	Method of counting
NaOH	$NaHCO_3$	Use dioxane base or ethanolic mixtures for counting aqueous solutions of the bicarbonate
$BaSO_4$	$BaCO_3$	Use such agents as SiO_2 to suspend the insoluble carbonate, which has to be finally divided to minimize self-absorption
Hyamine (Rohm & Haas, Inc.)	$R_1R_2R_3R_4{-}NCO_3$	Dissolve in methanol and add to aromatic base liquid scintillator. Impurities lead to quenching and chemiluminescence
Primene 81-R (Rohm & Haas, Inc.)	$R_1R_2R_3{-}N^+HCO_3^-$	Dissolve in alcohol and add to aromatic base liquid scintillator. Some question as to quantitative absorption of CO_2
Ethanolamine	$HOCH_2CH_2{-}NHCO_2$	Dissolve in methanol and add to toluene base liquid scintillator. Low quench, ^{14}C efficiency $\geq 85\%$, quantitative CO_2 absorption[a]
Phenethylamine	$-CH_2{-}CH_2{-}NHCO_2$	Dissolve in methanol and add to toluene base liquid scintillator. Has to be purified because of impurity quench. ^{14}C efficiency $\geq 85\%$. One investigation stated that a double salt is actually formed[b]
	$-CH_2{-}CH_2{-}NH_3^+$	
	$-CH_2{-}CH_2{-}NCO_2^-$	

[a] H. Jeffrey and J. Alvarez, *Anal. Chem.* **33**, 612 (1961).
[b] See F. W. Woeller, *Anal. Biochem.* **2**, 508 (1962).

collected, provided the tube is not blocked by the frozen water. This is very important in cases where large samples need to be combusted in order to get enough sample to be able to measure the radioactivity, i.e., low specific activity samples. Several samples can be burned and the water vapor trapped in the same trap. The whole sample can then be added to a liquid scintillator for counting.

In the case of ^{14}C the amount of sample that can be burned will be determined by the amount of trapping agent which can be incorporated in the scintillator solution. Since the phenethylamine or ethanolamine carbamates are sparingly soluble, appropriate amounts of methanol are necessary to have a homogeneous solution. If the sample is too large, the necessary amount of methanol will produce large amounts of quench. Assuming that at least 2 moles of the amine is necessary to react with 1 mole of CO_2, it is possible to calculate the sample size that can be combusted.

Example. If 400 mg sucrose, $C_{12}H_{22}O_{11}$, is combusted, how much ethanolamine or phenethylamine is required to trap quantitatively all of the CO_2 produced? Assume that it requires 2 mole of the amine to react with one mole of CO_2.

The physical data needed for the calculation include:

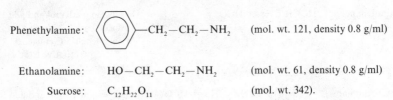

Phenethylamine: —CH_2—CH_2—NH_2 (mol. wt. 121, density 0.8 g/ml)

Ethanolamine: HO—CH_2—CH_2—NH_2 (mol. wt. 61, density 0.8 g/ml)

Sucrose: $C_{12}H_{22}O_{11}$ (mol. wt. 342).

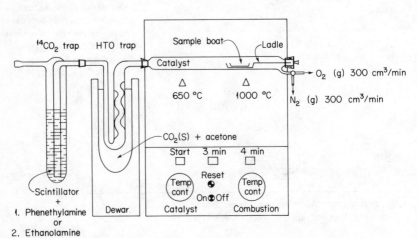

Fig. VII-3. Harvey biological material oxidizer. Dimensions: height, 20 in.; width, 15 in.; depth, 13 in.; weight, 30 lb.

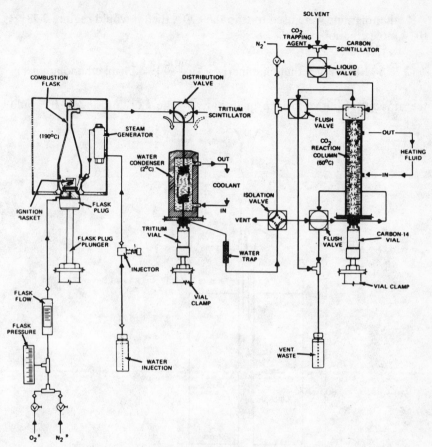

Fig. VII-4. Packard Model 305 Sample Oxidizer, flow diagram. (Asterisk (*) indicates from pressurized source.)

The complete combustion of one mole of sucrose will produce 12 mole of CO_2. Thus 400 mg sucrose will produce:

$$\left(\frac{0.400 \text{ g}}{342 \text{ g/mole}}\right)(12 \text{ mole } CO_2/\text{mole sucrose}) = 1.39 \times 10^{-2} \text{ mole } CO_2.$$

If phenethylamine were used to trap the CO_2, then

$$(1.39 \times 10^{-2} \text{ mole } CO_2)(2 \text{ mole phenethylamine/mole } CO_2)$$
$$= 2.78 \times 10^{-2} \text{ mole phenethylamine},$$

or

$$(2.78 \times 10^{-2} \text{ mole phenethylamine})\frac{121 \text{ g/mole}}{0.8 \text{ g/ml}} = 4.2 \text{ ml phenethylamine},$$

the amount required to trap quantitatively the CO_2 produced.

If ethanolamine were used to trap the CO_2, then it would require 2.78×10^{-2} mole ethanolamine:

$$(2.78 \times 10^{-2} \text{ mole ethanolamine})\left(\frac{61 \text{ g/mole}}{0.8 \text{ g/ml}}\right) = 2.1 \text{ ml ethanolamine},$$

the amount required to trap quantitatively the CO_2 produced. It would

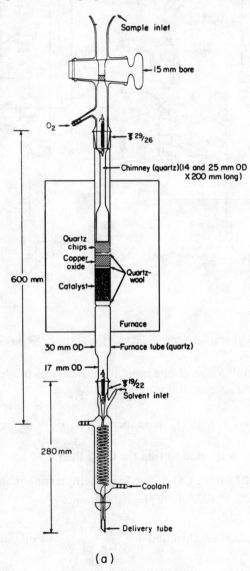

Sample inlet

15 mm bore

O_2

$\bar{\mathbb{Z}}\,^{29}/_{26}$

Chimney (quartz)(14 and 25 mm OD × 200 mm long)

Quartz chips

Copper oxide

Catalyst

Quartz-wool

600 mm

Furnace

30 mm OD — Furnace tube (quartz)

17 mm OD

$\bar{\mathbb{Z}}\,^{19}/_{22}$

Solvent inlet

280 mm

Coolant

Delivery tube

(a)

require twice the amount of phenethylamine to absorb the same amount of CO_2 as that absorbed by ethanolamine.

From the example, it is evident that for equal volumes, ethanolamine will trap twice as much CO_2 as phenethylamine. For large samples, it is probably necessary to use ethanolamine; however, it should be remembered that the carbonate of ethanolamine may not be as stable as the carbonate of phenethylamine. Elevated temperatures should be avoided. This will be critical if the warm gases from the combustion tube are introduced directly into the amine trapping solution.

Besides eliminating quenching, the combustion of the sample will lead to the complete separation of the ^{14}C and ^{3}H activities. Thus dual counting in a single sample is no longer necessary. This is very important for samples in which the ratio of ^{14}C to ^{3}H varies widely. In the combustion instruments the oxygen gas serves not only to support the combustion but to sweep the

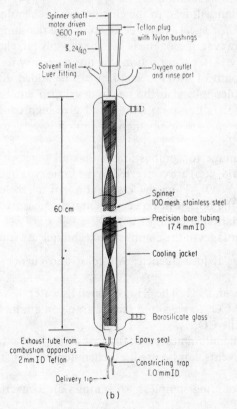

(b)

Fig. VII-5. (a) Peterson tritium combustion apparatus. (b) Peterson carbon–tritium absorption apparatus.

gases through the catalyst region and into the traps. Three types of commercial combustion apparatus are shown in Figs. VII-3–VII-5. The water vapor, which contains the ^3H, will be condensed in the cold trap, but the CO_2 will not condense. The CO_2, which contains the ^{14}C, will react with the amine in the second trap, and thus the ^3H and ^{14}C are separated. The H_2O can be dissolved in any of a number of water counting liquid scintillator solutions. The carbamate–methanol solution can either be assayed or completely dissolved in a liquid scintillator solution for counting.

The size of sample that can be burned in these combustion systems is to a great extent determined by the amount of CO_2 generated. Normally, if all of the carbamate–methanol solution is to be counted, the limit is about 400 mg of an organic-type compound. This amount of CO_2 requires about 2 ml of ethanolamine (4 ml of phenethylamine) and at least about twice as much methanol to prevent precipitation of the carbamate when diluted by the liquid scintillation solution.

Also, combustion will lead to the elimination of chemiluminescence, due to the sample and/or agents required to solubilize the sample in the scintillator solvent. However, if dioxane is used as the solvent, chemiluminescence is still quite likely. After combustion it is not necessary to use dioxane solvents. If the amount of water is small, it can be added directly to toluene with the aid of alcohol. Also the newly developed emulsion systems will easily handle larger amounts of water and give high counting efficiencies (up to 45%). The carbamate–methanol solution is very soluble in toluene base liquid scintillators.

Another advantage to combustion is that all types of samples (brain, liver, muscle, blood, etc.) are converted to the same type of counting sample, i.e., ^3HHO and $^{14}CO_2$. Thus the distribution of a labeled compound in various parts of the body can be very accurately measured by combustion and liquid scintillation counting.

The main advantages of the combustion technique are summarized below:

1. Eliminate or reduce quench: All color and chemical quenchers are eliminated.

2. Separate ^3H and ^{14}C: The ^3H is trapped as water in a cold trap, which allows the ^{14}C as CO_2 to pass to a second trap containing an amine, which chemically fixes the CO_2.

3. Eliminate chemiluminescence: Because all impurities are also oxidized, and aromatic solvents can be used, all chemiluminescence should be eliminated.

4. Common counting samples: All samples are converted to water and CO_2.

References

1. H. I. Jacobson, G. N. Gupta, C. Fernandez, S. Hennix, and E. V. Jensen, *Arch. Biochem. Biophys.* **86,** 89 (1960).
2. D. L. Buchanan and B. J. Corcoran, *Anal. Chem.* **31,** 1635 (1959).
3. W. J. Kirsten and M. E. Carlsson, *Microchem. J.* **4,** 3 (1960).
4. W. Schonigen, *Mikrochim. Acta* **1,** 23 (1955).
5. F. Kalbner and J. Rutschmann, *Helv. Chim. Acta* **44,** 1956 (1961).
6. V. T. Oliverio, C. Denham, and J. D. Davidson, *Anal. Biochem* **4,** 188 (1962).
7. J. D. Davidson and V. T. Oliverio, *in* "Advances in Tracer Methodology" (S. Rothschild, ed.), Vol. 4, p. 67. Plenum, New York, 1968.
8. R. G. Kelly, E. A. Pects, S. Gordon, and D. A. Buyske, *Anal. Biochem.* **2,** 267 (1961).
9. H. E. Dobbs, *Anal. Chem.* **35,** 783 (1963).
10. H. Jeffrey and J. Alvarez, *Anal. Chem.* **33,** 612 (1961).
11. H. G. Burns and H. Glass, *Int. J. Appl. Radiat. Isotop.* **14,** 627 (1963).
12. A. S. McFarlane and K. Murray, *Anal. Biochem.* **6,** 284 (1963).
13. S. Tsurufuji and J. Takashashi, *Radioisotopes* **14,** 146 (1965).
14. S. Von Schuching and C. W. Karickhoff, *Anal. Biochem.* **5,** 93 (1962).
15. H. P. Baden, *Anal. Chem.* **36,** 960 (1964).
16. C. F. Baxter and I. Senoner, *Anal. Biochem.* **7,** 55 (1964).
17. W. D. Conway, A. J. Grace, and J. E. Rogers, *Anal. Biochem.* **14,** 491 (1966).
18. R. E. Ober, A. R. Hansen, D. Mourer, J. Baukema, and G. W. Gwynn, *Int. J. Appl. Radiat. Isotop.* **20,** 703 (1969).
19. J. D. Davidson, V. T. Oliverio, and J. I. Peterson, "The Current Status of Liquid Scintillation Counting" (E. D. Bramsome. Jr.. ed.). p. 222. Grune & Stratton, New York, 1970.
20. L. A. Ford, *J. Ass. Off. Anal. Chem.* **53,** 86 (1970).
21. M. M. Griffiths and A. Mallinson, *Anal. Biochem.* **22,** 465 (1968).
22. W. E. Kisieleski, Private communication. Biol. and Med. Div., Argonne Nat. Lab., Argonne, Illinois, 1969.
23. G. N. Gupta, *Anal. Chem.* **38,** (1966).
24. G. N. Gupta, *in* "Organic Scintillators and Liquid Scintillation Counting" (D. L. Horrocks and C. T. Peng, eds.), pp. 747, 753. Academic Press, New York, 1971.
25. N. Kaartinen, Packard Tech. Bull. No. 18. Packard Instrum. Co., Downers Grove, Illinois, 1969.
26. J. I. Peterson, F. Wagner, S. Siegel, and W. Nixon, *Anal. Biochem.* **31,** 189 (1969).
27. J. I. Peterson, *Anal. Biochem.* **31,** 204 (1969).
28. D. W. Sher, N. Kaartinen, L. J. Everett, and V. Justes, Jr., *in* "Organic Scintillators and Liquid Scintillation Counting" (D. L. Horrocks and C. T. Peng, eds.), p. 849. Academic Press, New York, 1971.
29. T. R. Tyler, A. R. Reich, and C. Rosenblum, *in* "Organic Scintillators and Liquid Scintillation Counting" (D. L. Horrocks and C. T. Peng, eds.), p. 869. Academic Press, New York, 1971.

CHAPTER VIII

COUNTING VIALS

The vial serves the purpose of holding the sample and liquid scintillation solution. The vial should be transparent to the light scintillations produced in the liquid scintillation system, and should also be able to contain the sample and liquid scintillation solvents. There are several types of vials available commercially, each of which has certain advantages for specific needs: low background, high transmission, low cost, etc. Several of the commonly used types of vials are listed in Table VIII-1.

Since the common scintillator solutes emit photons of energy corresponding to wavelengths between 300 and 500 nm, it is required that the vial be transparent in that region. Common glass does not have a high transmission over this entire wavelength range. Figure VIII-1 shows the transmission of a borosilicate glass (i.e., Pyrex) and quartz.

It is evident from these data that some of the photons from PPO will be absorbed by borosilicate glasses but not by quartz. Thus use of a wave shifter, like M_2-POPOP, should increase the number of photons per unit excitation energy which escape from a borosilicate counting vial. The use

Table VIII-1

Types of liquid scintillation counting vials

Type	Main advantage	Main disadvantage
Glass (borosilicate, soda-lime)	Low cost	High background, absorption in near uv
Low K glass	Reduced background from ^{40}K	Absorption in near uv, expense
Quartz	Increased transmission in the uv	Expense
Plastic (i.e., polyethylene)	Low background, low cost, higher efficiency	Some solvents lost
Nylon	Low background	Cost, solvent loss
Teflon	Low background, can be re-used, high efficiency	Cost, has to be machined

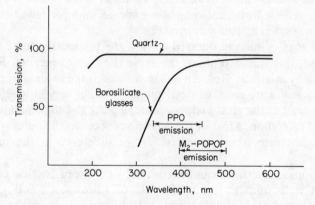

Fig. VIII-1. Transmission of quartz and borosilicates as a function of wavelength. Emission bands of some common scintillators are shown for comparison.

of quartz vials will not require a wave shifter to maximize the number of photons that will be transmitted through the vial walls. It should also be noted that the same problem is associated with the face of the multiplier phototube. If quartz-faced tubes are employed, it is not necessary to use a wave shifter. If borosilicate glass is used for the MPT face, an increased pulse height may be obtained by the use of a wave shifter.

The plastic vials, i.e., polyethylene, nylon, Teflon, give somewhat higher counting efficiencies for identical ^3H-containing samples. This higher efficiency may be the result of several factors, among which are

1. increased transmission of light,
2. creation of a more diffuse photon burst, which will interact with a larger area of the MPT (see Chapter IV on MPT "hot spots"),
3. reduction of light trapping in vial walls.

A series of stock Pyrex vials has been checked for transmission of light of wavelength 3500–5000 Å. Since the thickness of the vial walls varied, the light transmission was not always the same (1). In fact, the transmission varied from 60 to 90%, but it was improved by grinding the face until it was very thin. The transmission improved as the face became thinner.

Another factor, which should be considered in the selection of vials, is that light can be trapped within the walls of clear glass vials by total internal reflection. If the light photons hit the glass surface at an angle greater than the critical angle, those photons will be reflected. These reflected photons may reenter the scintillator solution and be subjected to further probable absorption by components in the solution (sample, impurities, solvents, or solutes themselves), which will result in a decrease in the photon yield. The reflected photons may also undergo multiple reflections within the glass walls, finally being trapped (2).

If light that is initially directed toward the photocathode of the MPT is reflected, this would produce the same effect as quenching. Reflection occurs at any interface. Reflections are reduced if the indexes of refraction of the two media are equal or nearly equal. There is very little reflection at the inside wall of the glass vial, because the glass and toluene have similar indexes of refraction. Most of the reflections occur at the outer wall at the glass–air interface where there is a large difference in the indexes of refraction.

Reflections at interfaces can be reduced by the use of "optical coupling." Common techniques might involve use of silicone oils and greases to couple surfaces or use of light pipes (2, 3). Filling the area between the vial and the faces of the multiplier phototubes with silicone oil, mineral oil, or glycerin has been attempted at various times. However, these lead to rather undesirable effects: films on the bottles which collect dust, oil drippings in other parts of the counter, and contamination of the interfacing medium, which can lead to reduction of photon transmission.

Photons which strike the outside walls of the vial will undergo either transmission with refraction, or reflection. What happens to the reflected photons will depend on the shape of the vial and nature of the scintillation solution. Some of the photons can be trapped inside the vial walls by repeated reflections on the outside and inside surfaces of the vial wall. For surfaces between two media of different indexes of refraction the critical angle is defined as the angle at which photons striking the interface with a greater angle, will be reflected. For the interface between glass and air (see Fig. VIII-2a):

$$\sin \gamma_0 = (\text{refractive index of air})/(\text{refractive index of glass}) = 1.0/1.5$$

or

$$\gamma_0 = 42°.$$

Thus all photons incident upon the glass surface with an angle greater than 42° will be reflected, and those with a smaller angle will be transmitted. At the interface between the toluene and glass (see Fig. VIII-2b):

$$\sin \gamma_0 = (\text{refractive index of toluene})/(\text{refractive index of glass}) = 1.5/1.5$$

or

$$\gamma_0 = 90°.$$

This states that essentially all photons will be transmitted across the interface without reflection. Since the refractive index varies with the wavelength of the photons, the angle of reflection will be a variable over the possible wavelength values dependent on the solute emission.

In certain cases, depending on the geometry of the vial, the reflected photons can be trapped inside the vial walls by multiple reflections until the photons are finally trapped. This trapping geometry can be interrupted. The interruptions will cause a change in the direction of the photons, and if the new angle is less than the angle of reflection, the photons will escape from the vial. Gordon and Curtis (4) studied methods of altering the symmetry by (1) introducing white cotton fabric into the scintillation solution, and (2) sandblasting the walls of the vials. Table VIII-2 shows the results they obtained.

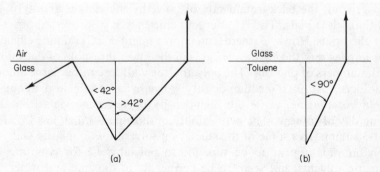

Fig. VIII-2. Illustration of the critical angle of photons at an (a) air–glass and (b) toluene–glass interface.

Table VIII-2

Effect of vial treatment on tritium counting efficiency[a]

Treatment	³H efficiency (%)
Clear glass vial	39.5
Clear glass vial + cotton fabric	44.8
Sandblast inside of vial	40.0
Sandblast outside of vial	46.0
Sandblast outside + cotton fabric	43.2
Sandblast outside in strips	45.3

[a] See B. E. Gordon and R. M. Curtis, *Anal. Chem.* **40**, 1486 (1968).

Sandblasting the inside walls of the glass vials had little effect, because there are essentially no reflections between the glass and toluene. The sandblasting on the outside had the greatest effect, because symmetry was altered at the interface of different indexes of refraction.

The same effect of increased efficiency for ³H counting has been noted with plastic vials. Since most plastic vials are milky, it is believed that very little light is trapped in the vial walls. There are also several other distinct advantages to the use of plastic vials, including (1) lower background, (2) higher photon yield (diffuse vial walls), (3) lower phosphorescence, and (4) their low cost (discarded after use). However, one major drawback to plastic vials is that many solvents used in liquid scintillation diffuse through the plastic vials. This loss of solvent leads to many undesirable effects, including (1) increased concentration of fluors and sample, (2) if quenchers are present in sample, there will be increased quenching, (3) change in size of plastic walls upon absorption of solvent, and (4) vial size change leads to malfunction of automatic sample changers.

In general, the background rate of polyethylene vials is about 10–20% less than glass vials. The ³H counting efficiency is improved about 3–5% over glass vials. However, there is no improvement in ¹⁴C counting efficiency. This implies that the greatest effect is in the lower threshold of detection of small numbers of photons. The polyethylene vials lose rather large amounts of toluene. At 0°C medium-density polyethylene vials lose about 100 mg/day of toluene, while the high-density polyethylene vials lose about 35 mg/day of toluene. The permeability of dioxane is much less (5). Table VIII-3 summarizes some of the studies on solvent loss in plastic vials.

Nylon vials appear to be superior to polyethylene for counting with aromatic solvents, but seem to be permeable to dioxane and water. The Teflon vials, custom-made by Calf (6, 7), seem to be the best but are very expensive and are not supplied in any large quantities or by any of the commercial concerns. The Teflon vials can be reused.

Table VIII-3

Measurement of solvent loss in various types of plastic counting vials[a]

		Volume	Loss in mg/1 day			
Vials	Supplier	(ml)	p-Xylene	Toluene	Dioxane	H$_2$O
Poly-	Nuclear-	25	128	187	4.1	0.5
ethylene	Chicago					
	Packard	25	120	192	3.2	0.1
	Beckman	20	77	101	1.2	0.3
Nylon	Nuclear-	25	5.3	10	15	34
	Chicago					
Teflon	G. Calf[b]	20	None	None	None	None
	custom-made					

[a] See R. Liberman and A. A. Moghissi, *Int. J. Appl. Radiat. Isotop.* **21**, 319 (1970).
[b] See G. E. Calf, *Int. J. Appl. Radiat. Isotop.* **20**, 611 (1969).

Calf (6) studied the background of several different types of vials and the ^{3}H counting efficiency with two different scintillator solutions in the vials. One main advantage of Teflon vials is the very low background rate. Table VIII-4 summarizes the data.

Paix (8) investigated the background of two types of glass vials compared to a polyethylene vial. The polyethylene vial had a background rate in the ^{3}H counting channel of about one-half that for glass vials. But he also noted that there is a 30–40% variation in the observed background rate of glass vials. Several vials were crushed, and 500-g samples of the powdered glass

Table VIII-4

Background and tritium counting efficiency for different types of counting vials[a]

	Aqueous samples[b]			o-Xylene samples[c]		
Vial type	Efficiency (%)	Background (cpm)	E^2/B	Efficiency (%)	Background (cpm)	E^2/B
Low K glass	31.5	25.5	39	56.8	28.1	115
Quartz	31.5	11.5	86	59.0	18.4	189
Polyethylene	32.8	8.6	125	60.1	13.9	260
Teflon	31.7	7.0	144	57.6	11.6	286

[a] See G. E. Calf, *Int. J. Appl. Radiat. Isotop.* **20**, 611 (1969).
[b] Three grams of ^{3}H$_2$O and 15 ml of a dioxane scintillator solution. A wave shifter was used, which explains the same efficiency in glass and quartz vials.

Fifteen milliliters of o-xylene with 4.0 g/liter of PPO and small amount of o-xylene-^{3}H. No wave shifter present; note the different efficiency in glass and quartz.

were counted in a low-level gamma spectrometer with 33% geometry. The predominant γ rays were due to ^{226}Ra and its daughters. ^{40}K was absent, to the limit of the measurement, and ^{7}Be was present in the American-made vials. Table VIII-5 gives some typical mean background rates for counting vials.

Table VIII-5

Mean background rates for different types of counting vials [a]

Vial	Number of vials counted	Mean background cpm in ^{3}H channel
Polyethylene	111	25.05
	10	21.10
Wheaton glass (20 ml)	10	39.7
	10	40.6
	10	37.8
	10	48.9
	10	53.5
	1[b]	36.4
Australian glass (20 ml)	9	55.4
	10	54.7
	10	43.5
	10	43.8
	10	46.2
	1[b]	41.2

[a] See D. Paix, *Int. J. Appl. Radiat. Isotop.* **19**, 162 (1968).
[b] Single vial counted eight times, with unloading between counts.

Many types of counting vials exhibit an emission of light which is not the result of excitations from radioactivity. These are usually some type of photoluminescence, phosphorescence, and/or fluorescence, excited by light, usually sunlight or room light. The dissipation of this excitation energy usually has two time constants, one which decays very rapidly and another which decays much more slowly. The rapid component may be completely gone in a time of minutes in some cases, to hours in others. The slow component may still contribute to the measured cpm after days (9).

Different types of vials have different amounts of these photoluminescence emissions. Most plastic vials appear to have much less photoluminescence than glass vials (Fig. VIII-3). And more important, the photoluminescence seems to be predominantly the shorter-lived variety in the plastic vials. In one experiment glass and plastic vials were placed in the direct sunlight for 4 hr (9). The deexcitation was not complete, even after six weeks.

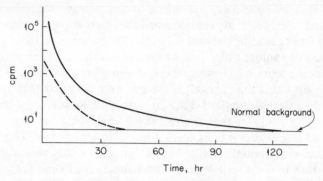

Fig. VIII-3. Vials irradiated for 10 min with sunlight, placed in counter and kept in darkness between counting times: —— glass vials; ———— polyethylene vials.

The automatic sample changers have introduced another source of background counts. The continuous motion of the vials, by sliding, leads to buildup of static charge on vials (9). The charge buildup seems to be most troublesome with the plastic vials, which seem to retain the static electricity and release it in a sudden discharge, which results in a light flash. It is very hard to predict or determine quantitatively the decay time of the static electricity.

The light flashes from static electricity discharges are usually weak and correspond to the scintillations from low-energy electrons. In most experiments the static electricity is a problem only for systems with very high efficiency. Sometimes the static electricity can produce excitations in the liquid scintillation solution itself. In one study of static electricity effects, it was shown that prior treatment of the plastic vials with an antistatic compound eliminated any noticeable static electricity without decreasing the counting efficiency (9).

When high-energy radiation is being measured, a wall effect is often observed. The wall effect results from the range of the ionizing radiation being such that a fraction of the particles reaches the wall of the vial before the total energy is expended (10). This results in a photon yield which is less than the photon yield corresponding to completely stopping the ionizing particle in the scintillation solution. This gives the same result as chemical or color quenching, except that the higher-energy particles are affected rather than the lower-energy particles.

Recently it has been observed (11–14) that not only does the scintillator solvent diffuse into and through the walls of plastic (high-density polyethylene) vials, but the scintillator solutes diffuse into the walls also. There are two adverse results of this diffusion: First, it has been observed that the plastic, when impregnated with scintillator solutes, scintillates. As a result of this, many pulses, equivalent to low-energy electron excitations in the liquid scintillator, are created when the external γ-ray source (external standard) irradiates the vial and solution. The number of these pulses depends upon the amount of solute which has diffused into the plastic. Since the diffusion occurs over several days, the number of such pulses will be increasing over this time period. If the counting channels that are used to measure the external standard channels ratio (ESCR) value, for monitoring quench, include any of these pulses, the ESCR value has been observed to drift, and if the cause is not known it can be interpreted as a change in the quench level of the sample. Figure VIII-4 shows the ESCR values of a scintillator solution in a polyethylene vial as a function of time for two ESCR counting channels selections. Curve (a) was obtained with the ESCR counting channels set to exclude all pulses which originate in the plastic vial walls. Curve (b) was obtained when one of the ESCR counting channels was set to include pulses which originated in the plastic vial walls. The sample, in fact, had no quench change over the total time of the measurement.

The second adverse effect of the solute diffusion was an increase in measured background with time. The background was mostly due to additional

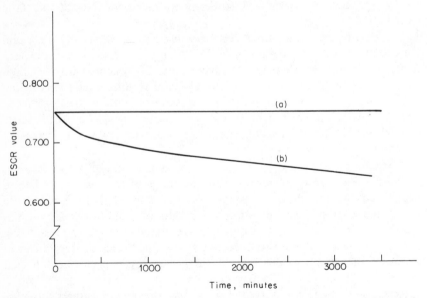

Fig. VIII-4.   ESCR value as function of time that a scintillator solution was stored in a high-density polyethylene vial. Curve (a) was obtained when ESCR counting channels were selected which excluded all pulses originating in the plastic walls of the counting vial. Curve (b) was obtained when one of the ESCR counting channels included pulses originating in the plastic walls.

pulses of low amplitude which were registered in the tritium counting channel. The background rate in a tritium count channel was observed to increase from 12 cpm at the time of addition of the scintillator solution to the polyethylene vial, to over 21 cpm after 6 days storage of the scintillator solution in the vial.

References

1. D. L. Horrocks, *Rev. Sci. Instrum.* **35,** 334 (1964).
2. R. K. Swank, *in* "Liquid Scintillation Counting" (C. G. Bell and F. N. Hayes, eds.), p. 23. Pergamon Press, Oxford, 1958.
3. D. L. Horrocks and M. H. Studier, *Anal. Chem.* **30,** 1747 (1958).
4. B. E. Gordon and R. M. Curtis, *Anal. Chem.* **40,** 1486 (1968).
5. R. Liberman and A. A. Moghissi, *Int. J. Appl. Radiat. Isotop.* **21,** 319 (1970).
6. G. E. Calf, *Int. J. Appl. Radiat. Isotop.* **20,** 611 (1969).
7. G. E. Calf, *in* "Organic Scintillators and Liquid Scintillation Counting" (D. L. Horrocks and C. T. Peng, eds.), p. 719. Academic Press, New York, 1971.
8. D. Paix, *Int. J. Appl. Radiat. Isotop.* **19,** 162 (1968).
9. A. A. Moghissi, H. L. Kelley, J. E. Regnier, and M. W. Carter, *Int. J. Appl. Radiat. Isotop.* **20,** 145 (1969).
10. D. L. Horrocks and M. H. Studier, *Anal. Chem.* **36,** 2077 (1964).
11. E. B. Mueller, *in* "The Current Status of Liquid Scintillation Counting" (E. D. Bransome, Jr., ed.), p. 181. Grune & Stratton, New York, 1970.
12. K. J. Johanson and H. Lundquist, *Anal. Biochem.* **50,** 47 (1972).
13. B. H. Laney, *in* "Tritium" (A. A. Moghissi and M. W. Carter, eds.), p. 156. Messanger Graphics, Las Vegas, Nevada, 1973.
14. D. L. Horrocks, to be published, 1974.

CHAPTER IX

BACKGROUND

There are many sources of background events in liquid scintillation counting. The main division of these sources is into those produced in the liquid scintillator solution and those which result from events that have no connection with the liquid scintillator solution. Table IX-1 summarizes many of the common sources.

Noise

Operation at high sensitivity of the multiplier phototube for measurement of the very low photon yields from low-energy beta particles is also responsible for the very high spontaneous emission of electrons from the photocathode and dynodes. These spontaneous events are commonly called MPT noise. The noise levels are characteristic of each individual MPT. Some tubes have very low levels of noise, whereas others with the same

Table IX-1

Sources of background

Source	Contributors
Liquid scintillator materials	Natural radioactivity in the materials which constitute the liquid scintillator Chemiluminescence and phosphorescence of certain solvents enhanced by the presence of the solutes
Sample	Natural radioactivity in the sample which may be the same or different from the nuclide to be assayed Contamination with the same or other radioactive material Chemiluminescence and phosphorescence produced by the sample or impurities in the sample
Vial	Natural radioactivity in the vial walls or cap Cosmic-ray-induced background—Cerenkov and secondary electrons and γ rays Chemiluminescence and phosphorescence produced by sun light or impurities on the vial walls Static charge buildup during movement in the sample changer
MPT	Natural radioactivity in materials which make up the MPT Cosmic rays which produce Cerenkov radiation, secondary electrons, and γ rays Thermionic and secondary electron emission from photocathode and dynodes—in coincidence systems this is mostly eliminated, because of its randomness Cross talk from electric discharges and/or Cerenkov radiation Afterpulses
Other radioactive sources	Radioactive sources (usually γ rays) in the area of the liquid scintillation counter. The movement of these sources can be very detrimental, because it will lead to changes in the background level

sensitivity will have very high levels of noise. The noise is a random process. The use of two MPTs in coincidence will eliminate nearly all background due to the random noise. The possibility of a chance coincidence of noise pulses in the two MPTs will be a function of the random rate of each tube (N_1 and N_2) and the resolving time (τ) of the coincidence circuit. The number of coincidence noise pulses N_c is given by

$$N_c = 2\tau N_1 N_2.$$

For modern commercial liquid scintillation the resolution time is between 20 and 30 nsec. If two MPTs with random noise levels of 10,000 cpm are used in a coincidence system with $\tau = 25$ nsec, the chance coincidence rate will be

$$N_c = 2\left(\frac{25}{60} \times 10^{-9} \text{ min}\right)(10,000)(10,000) = 0.08/\text{min}.$$

The contribution of tube noise to background is negligible in the modern liquid scintillation counters unless the MPT noise level of each tube exceeds approximately 30,000/min. (This will give $N_c \sim 1$ cpm.)

The actual chance coincidence rate of a particular counter can be obtained by blocking the view between the two MPTs with a black barrier. Most commercial instruments will give chance coincidence rates less than 1 cpm. Since the noise pulses are equivalent to less than two or three photoelectrons per pulse, the background due to random noise coincidence will be almost entirely in the ^{3}H counting channel.

The noise rate of the older MPTs, used in the earlier liquid scintillation counters, was very high. The noise rate is drastically reduced by decreasing the temperature. For this reason, most early commercial liquid scintillation counters were refrigerated. With the availability of the newly developed bialkali MPTs, the need for refrigeration, to reduce tube noise, is no longer necessary. (There may be other reasons for temperature control systems.) The noise level of the bialkali MPTs at room temperature is, on the average, even lower than the noise level of the older MPTs at $-20°$C.

Cross Talk and Gas Discharges

The term cross talk is used to describe the result of a light-producing event occurring in one of the two MPTs which is seen at the same time by the second MPT. This process results in a coincidence pulse. The light-producing event can be

(a) an electrical discharge,
(b) Cerenkov radiation from cosmic radiation,
(c) gas discharge (residual gas in MPT), or
(d) natural radioactivity in the construction material of the MPT.

The path of photons between the two MPTs can be through the sample itself, piped by the vial walls, or around the vial in a light guide or reflected by the optical system of the sample counting chamber.

Since the initial light-producing event occurs in one of the MPTs, the pulse amplitude in that MPT will be much larger than any coincident pulse produced in the other MPT. However, the pulse summation circuit will give a single pulse whose amplitude will be the sum of the two coincident pulses. Thus, most cross-talk pulses will have amplitudes which correspond to pulses from electrons which have relatively high energy, i.e., energies greater than the highest energy betas of ^{3}H decay. Thus, elimination of cross talk should give a greater improvement in ^{14}C background than in ^{3}H background (1).

Gas discharges result from residual gases, which remain after evacuation of the MPT. These gases can cause two types of background, afterpulses and cross talk (2). Afterpulses result from ionization of the gases during the amplification of high-energy-produced pulses. The positive gas ions will accelerate toward the photocathode and upon striking it will produce a second pulse, which would occur outside the resolving time of the electronic integration. The delay time will be determined by the time required for the gas ions to travel from their source to the photocathode. If there is no coincident afterpulse in the second MPT, the afterpulse will go undetected. However, because the afterpulses are the result of true coincident events, there is a high probability of coincident afterpulses if there are residual gases in both MPTs (3, 4).

The gases can also produce a discharge as a result of the high voltages in the MPT. These discharges will not be coincident unless the discharge gives rise to enough photons of the proper energy, which causes the photocathodes of both MPTs to emit photoelectrons coincidently.

Cerenkov and Cosmic Radiation

High-energy radiations can interact with materials in the vicinity of the liquid scintillator and MPTs to produce background counts. These interactions result in the production of Cerenkov radiation, secondary electrons, and low-energy γ rays and X rays. The low-energy electrons normally contribute to the background only when they are produced in the liquid scintillator. The low-energy X rays have a greater range of influence because of their greater penetration. The shielding material is a prime source of these X rays because of the high cross section for the high-energy cosmic rays. Thus the lead shield will be a source of Pb X rays (75 and 12 keV).

A prime source of background is the Cerenkov radiation produced by the slowing down of high-energy electrons in the glass envelope of the MPT or the vial walls. Since the Cerenkov radiation is photons, and part of the photons are of the same energy as the scintillator photons, the MPTs will count the Cerenkov radiations with a good efficiency. It is difficult to reduce Cerenkov radiations, by such tricks as optical filters, without also absorbing photons from real events.

The low atomic numbers of the materials that make up the liquid scintillator reduce the probability of the scintillator's interaction with cosmic radiations. Table IX-2 shows the relative background rates under various

Table IX-2

Typical background rates for various counting conditions[a]

Condition	Typical background (cpm)
Black bottle	≤ 1.0
Empty chamber	12
Empty vial	28
Vial filled with H_2O	28
Vial filled with liquid scintillator	28

[a] Measured rates in a tritium counting window.

conditions. Filling the vial with water did not alter the background, nor did the background differ significantly between an empty vial and one filled with liquid scintillator solution.

Natural Radioactivity

Natural radioactivity of different kinds will be present in all the materials used to make the liquid scintillation system—even the scintillator solution itself. Table IX-3 lists some of the common types of natural radionuclides present in various parts of the liquid scintillation system.

It is of interest to consider the source of the solvents used for liquid scintillation counting. If these solvents are from sources which utilized materials that were living in the present era, the carbon and hydrogen would have ^{14}C and ^3H present in the same amount as is present in the environment. Table IX-4 lists the modern-day concentrations of ^{14}C and ^3H in the atmosphere and water. It is recommended that all solvents and solutes be derivatives of petroleum products which, because of their old age, have essentially no ^{14}C or ^3H present.

Example. What is the disintegration rate of the ^{14}C present in 3 ml ethanol made from modern plants? Assume the modern concentration of ^{14}C to be 14 dpm/g carbon. How would this affect the background count rate?

The weight of ethanol is 3 ml $\times$ 0.8 g/ml or 2.4 g. Ethanol is 52% by weight carbon (24/46 = 0.52). The weight of carbon in 3 ml ethanol is (2.4 g

Table IX-3

Sources of natural background in various parts of the scintillation counting system

Source	Types of natural radioactivity
Scintillator solution	^{3}H and ^{14}C from atmospheric equilibrium or contaminations during preparation and/or purification of solvents and solutes
Sample	Any radionuclide, depending on the nature of the sample
Vial	^{40}K, ^{232}Th and daughters, ^{7}Be, and other nuclides present in the materials used to make glass vials. Use of plastic vials eliminates this source
Chamber	^{226}Ra and other nuclides, which will depend on the material used for chamber walls. If a part is used for the optical light collection system, there may be natural activity present; i.e., use of $BaSO_4$ may introduce ^{226}Ra and other natural radioactivity associated with Ba in nature. Lead shielding should be from sources which have not been near a reactor
MPT	^{40}K, ^{232}Th and daughters, normal uranium, etc., which may be in the glass used to form the outside and face of the MPT. Also, there may be other radioactive nuclides in the construction materials of the dynodes and insulators

Table IX-4

Levels of ^{14}C and ^{3}H in living matter in equilibrium with the environment

Isotope	Modern-day levels		Typical amounts
^{14}C	14 dpm/g of C		10 ml of toluene would contain $\sim$230 dpm of ^{14}C
^{3}H	Rainwater	100–1000 TU	1 ml of H_2O contains 0.72–7.2 dpm
	Rivers and shallow lakes	10–1000 TU	1 ml of H_2O contains 0.072–7.2 dpm
	Underground lakes	0–1000 TU	1 ml of H_2O contains 0–7.2 dpm

ethanol) $\times$ 0.52 g carbon/g ethanol) or 1.25 g. Therefore, the ^{14}C dpm rate would be given by:

$$(14 \text{ dpm/g})(1.25 \text{ g}) = 17.5 \text{ dpm/3 ml ethanol.}$$

If the ^{14}C counting efficiency were 90% with the 3 ml ethanol in the scintillator solution cocktail, the background rate would be increased by:

$$(17.5)(0.90) = 15.5 \text{ cpm.}$$

Example. A wine was made with water which had a tritium content of 1000 tritium units (TU). What will the tritium disintegration rate be when the sealed wine is 7 years old?

One ml water contains

$$\left(\frac{1\ g}{18\ g/mole}\right)\left(\frac{2\ ^1H\ atoms}{molecule}\right)(6.02 \times 10^{23}\ molecule/mole)$$

$$= 6.7 \times 10^{22}\ ^1H\ atoms.$$

One ml water of 1000 TU level will contain

$$N = (6.7 \times 10^{22}\ ^1H\ atoms)\left(\frac{1000\ ^3H\ atoms}{10^{18}\ ^1H\ atoms}\right)$$

$$= 6.7 \times 10^7\ ^3H\ atoms/ml\ water.$$

The disintegration rate (in dpm) when the wine was prepared is given by:

$$\frac{dN}{dt} = \lambda N = \frac{0.693}{t_{1/2}}N,$$

where $t_{1/2}$ is the half-life of tritium, 12.24 yr or 6.45×10^6 min. Therefore, the decay rate is:

$$\frac{dN}{dt} = \frac{(0.693)(6.7 \times 10^7\ atoms)}{6.45 \times 10^6\ min} = 7.2\ dpm/ml.$$

At the end of 7 years, the tritium would have decayed for (7/12.24) or 0.57 of one half-life. The activity at any time A_t can be calculated from the activity at time zero A_0, by the equation

$$A_t = A_0 e^{-\lambda t} = A_0 e^{-(0.693)(t/t_{1/2})}$$
$$= 7.2\ \exp[-(0.693)(0.57)]$$
$$= (7.2)(0.674)$$
$$= 4.8\ dpm/ml\ water\ from\ the\ 7\text{-}yr\ old\ wine.$$

Chemiluminescence

Chemiluminescence (CL) is the emission of photons as the result of a chemical reaction. This reaction may involve impurities present in the scintillator solution, reagents used to solubilize the sample, or the sample itself, and will continue as long as the chemical reactants are present. The rate of the reaction will determine the number of the photons emitted per unit time. When the rate is rapid, the photon intensity is high, but the number will decrease very rapidly. On the other hand, if the rate is slow, the intensity will be low and photon emission will be long-lived. The rate of the reaction is also dependent on such things as concentration of the reactants, temperature, etc., which are discussed in detail in Chapter XII.

The major problem with chemiluminescence is that it will vary with time. Thus the background of a sample that has a very small amount of CL will appear to decrease over a period of time. The rate of the decrease will be determined by the rate of the chemical reaction, which produces the excited species which lead to CL. In those cases where the CL rate is very fast, it is possible to wait a period of time before measuring the background and/or the sample count rate.

In liquid scintillation counting, experience has shown certain types of systems which are much more likely to exhibit CL than others. Using solvents, solutes, samples, solubilizers, additives, and other chemicals of the highest purity has been demonstrated as one way to reduce and minimize the chance of CL. But even the highest purity of all chemicals used will not completely eliminate the possibility of CL occurring.

It is often very difficult to determine if a sample does have CL, especially if the intensity of the CL is very small. One suggested method involves a two-channel counting arrangement. Figure IX-1 shows typical pulse spectra for CL and a weak beta emitter, such as ^{3}H. Using two counting channels, A and B, the counting efficiency of the beta emitter is determined in each channel for a sample which does not have CL.

Then the unknown sample is counted and the counts in channels A and B measured. The dpm of the sample is calculated. If the dpm values for the two channels are the same, no CL is present. If different dpm values are obtained, it can be assumed that some other source of counts is present. If the dpm for channel A is greater than that for channel B, it could be the result of CL in the sample. Often, measurement at a later time, after the CL has decayed, will give different dpm values.

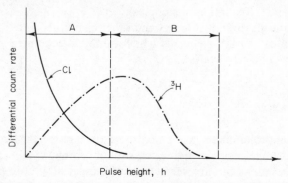

Fig. IX-1.   Pulse spectra for a chemiluminescence reaction and ^{3}H beta particles.

Example. The counting efficiency for ^3H in two channels is determined for a sample with no CL present:

$$\text{Eff}_A = 33\%, \qquad \text{Eff}_B = 17\%.$$

An unknown sample was counted, giving, after background subtraction:

$$\text{counts}(A) = 3700, \qquad \text{counts}(B) = 1720.$$

Was CL present in the sample?

The dpm of the sample calculated from the counting data was

$$\text{channel A dpm} = \frac{3700 \pm 121}{0.33} = 11{,}200 \pm 363$$

$$\text{channel B dpm} = \frac{1720 \pm 83}{0.17} = 10{,}100 \pm 490.$$

Therefore, it is likely that CL was present in the unknown sample.

Another method for the detection of CL in a sample is repeated counting of the sample over a period of hours, or even days. The sample count rate, provided the half-life is sufficiently long, will not change, but the CL contribution will decrease with time, and the combination of two-channel counting and repeated counts should be very sensitive for the detection of small amounts of CL.

Of course, the ideal sample counting system would be one which does not have CL. There are several methods for reduction or elimination of CL, some of which are:

1. storage of samples for long periods before counting,
2. storage of samples at elevated temperature (to increase reaction rate),
3. use of technique that eliminates the necessity of chemicals that are likely to cause CL (oxidation of samples),
4. special purification of solvents and other chemicals, and
5. addition of special chemicals which inhibit or compete with CL reactants to give non-CL products.

Photoluminescence

The term photoluminescence is used here to describe the production of photon-producing species by light, i.e., sunlight, room light, etc. Often the photoluminescent species are very long-lived, especially those produced in the vial walls and caps, since they are not subject to the normal deactivation processes that occur in the liquid phase via collision. This long-lived photo-luminescence is usually referred to as phosphorescence, although the strictest use of the term phosphorescence applies to the spontaneous emission from triplet excited states.

References

1. B. H. Laney, *in* "Organic Scintillators and Liquid Scintillation Counting" (D. L. Horrocks and C. T. Peng, eds.), p. 991. Academic Press, New York, 1971.
2. F. N. Hayes, R. D. Hiebert, and R. L. Schuck, *Science* **116**, 140 (1952).
3. D. L. Horrocks and M. H. Studier, *Anal. Chem.* **30**, 1747 (1958).
4. D. L. Horrocks and M. H. Studier, *Anal. Chem.* **33**, 615 (1961).

CHAPTER X

QUENCH CORRECTION METHODS*

Because of the quench effects of very small amounts of some materials, it is very difficult to prepare two identical counting samples, i.e., sample and liquid scintillation cocktail. The different amount of quench will be reflected in the counting efficiency. More quench will cause a lower counting efficiency, less quench a higher counting efficiency. The absolute change in counting efficiency will be greater as the energy range of the particles to be counted decreases. The same increase in quench will reduce the counting efficiency for ^3H more than for ^{14}C. Figure X-1 shows the counting efficiencies for ^3H and ^{14}C as a function of the quench as measured by the external standard channels ratio S which is described in detail later. A lower S value indicates a greater amount of quenching agent. From this plot it is readily observed that equal quench levels lead to a greater reduction of the ^3H counting efficiency than of the ^{14}C counting efficiency.

* Much of the data presented in this chapter were obtained using a Beckman LS-250 liquid scintillation counter and should serve only as an example. Similar data can be obtained on any of the commercially available instruments.

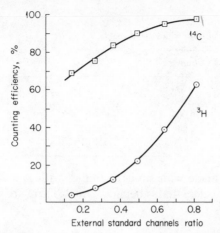

Fig. X-1. Counting efficiencies of ³H and ¹⁴C as a function of quench as monitored by the external standard channels ratio. Equal quench levels produce greater decrease in ³H efficiency than in ¹⁴C efficiency.

In order to measure accurately the amount of radioactive material in a liquid scintillator solution, it is necessary to know the real counting efficiency of that nuclide. Therefore, it is necessary to monitor the efficiency of the sample in the liquid scintillator solution. The actual dpm of a radioactive nuclide is obtained from the measured cpm by the equation:

$$dpm = cpm/Eff.$$

The accuracy of the determination of the sample dpm will depend on the accuracy of both the cpm value and the efficiency. The accuracy of the cpm is obtained from the total number of counts recorded. The variation is usually given as the sigma value σ, which is described in detail in Chapter XVIII:

$$\sigma = (\text{total counts recorded})^{1/2} = \sqrt{n}$$

$$\%\sigma = (\sqrt{n}/n) \times 100 = (1/\sqrt{n}) \times 100.$$

The greater the number of counts recorded, the smaller is the error. Therefore the length of time that a sample is counted is related to the accuracy of the cpm value. If a sample has 1000 cpm, the error of the measurement will vary with time of count as shown in Table X-1.

The second factor of variation in measurement of the actual amount of radioactivity in the sample is the uncertainty in the counting efficiency. The counting efficiency is usually determined by some method which relates

Table X-1

Variation of measurement error with time of count

Time (min)	Counts	σ (Counts)	σ (%)	2σ (%)
1	1000	31.6	3.2	6.4
5	5000	70.7	1.4	2.8
10	10,000	100.0	1.0	2.0
50	50,000	224.0	0.45	0.90
100	100,000	316.0	0.32	0.64

to measurements made with similar samples with known amounts of the same radioactive nuclide that is being counted. There are several techniques employed.

Quench Monitoring Methods

Internal Standard

The internal standard technique is probably the oldest method (1–4). It involves the addition of a known amount of the nuclide in high specific activity to the same sample that is being measured. The counting efficiency is then calculated from:

$$\text{Eff} = \frac{\text{cpm (standard + sample)} - \text{cpm (sample)}}{\text{dpm (standard)}}.$$

The cpm of the standard is usually many times greater than the cpm of the sample. Therefore, the statistical uncertainty of the sample cpm will not be important in the uncertainty of the efficiency.

The procedure for the internal standard method usually involves the following steps:

1. count the sample to obtain cpm (sample),
2. add small amount of high specific activity of same nuclide being measured in sample directly to the counting vial,
3. recount to obtain new cpm (standard + sample), and
4. calculate efficiency from equation.

This method has some advantages. It is rapid, assuming the added activity is high enough to obtain a statistically accurate value in a short time. The efficiency is measured in the actual sample, which eliminates the necessity to run a series of quenched standards. Different types of quenchers will not affect the validity of this method.

On the other hand, there are several major drawbacks to the use of this method. First, the sample cannot be recovered if the internal standard is identical to the sample. More important, the sample cannot be recounted to check its cpm value. There are certain hazards to opening up the vial and adding something extra to the scintillator solution. In adding the internal standard it is possible also to add extra quench, and in some instances (contaminated pipet) variable amounts of quench which lead to unreal, low counting efficiencies and thus to high calculated amounts of the radioactive nuclide in the sample. With refrigerated samples, opening the vial to add the internal standard could lead to condensation of moisture in the vial. Water is a very strong quencher. The amount of moisture will vary, depending on the relative humidity.

Sample Channels Ratio

A very good method of monitoring quench is called the "sample channels ratio (SCR) method" (5–7). Since most nuclides counted by liquid scintillation are beta emitters, the pulses produced as the result of the beta particles being stopped will have amplitudes varying from zero to some maximum. The distribution of pulse amplitudes will be similar (although not exact) to the beta-particle energy distribution. By monitoring the ratio of counts in two channels, it is possible to measure the amount of quench. The method does not require the addition of anything to the scintillator solution, and the sample can be recounted as often as needed. The method is limited to the level of radioactivity in the sample and does not give as reliable results if the quench is great.

The distribution of pulses for a beta-emitting nuclide in a liquid scintillator will be similar to that shown in Fig. X-2. The two counting channels can be selected in several ways. Three typical counting modes are given in Fig. X-3. This method requires a standard quench curve. A series of samples, as nearly identical as possible with the samples to be measured, are counted which have known amounts of the radioactive nuclide and increasing

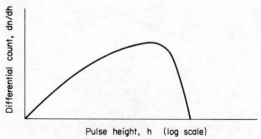

Pulse height, h (log scale)

Fig. X-2. Typical pulse spectrum of beta-emitting nuclide.

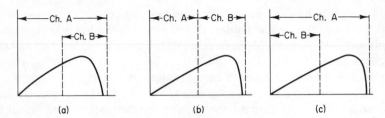

Fig. X-3.   Selections of counting channels for sample channels ratio method.

amounts of quenching agent. The counting efficiency and the sample channels ratio are measured for each quenched standard. The counting efficiency is plotted as a function of the sample channels ratio. Subsequently, an unknown sample is counted and its sample channels ratio measured. The counting efficiency for the unknown sample is obtained from the standard quench plot. Figure X-4 shows a typical plot for a ^{14}C sample quenched with increasing amounts of CH_3NO_2 (nitromethane), using a channel selection as shown in Fig. X-3a. Any sample which measures an SCR value of 0.50 will be counted with an efficiency of 60%. Figure X-5 shows the variety of plots that can be obtained for tritium by different selections of the "B" counting channel with the "A" counting channel set to cover the total pulse spectrum of an unquenched ^{3}H sample.

The SCR efficiency plot (slope, linearity, etc.) is dependent on the choice of the discriminator setting, which defines the counting channels. The choice of the discriminator setting will depend on the quench range to be covered. If the quench range is small, a good setting for the discriminators is that shown in Fig. X-6 for maximum change in the SCR with quench change. A small change in the amount of quench will make a large relative change in the counts in the B channel and thus a large change in SCR value

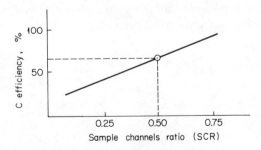

Fig. X-4.   Plot of sample channels ratio as ^{14}C counting efficiency using counting channels as shown in Fig. X-3a. Unknown samples with SCR of 0.50 will have 60% counting efficiency for ^{14}C.

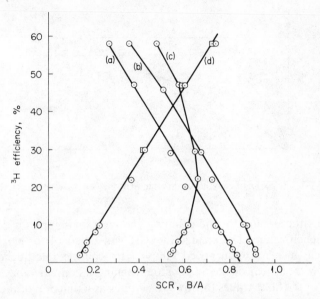

Fig. X-5. Plot for ^3H with SCR equal to counts in channel B divided by counts in channel A. Channel A is always 0–300. Channel B is (a) 0–90, (b) 0–110, (c) 50–150, and (d) 90–330.

which is defined as the ratio of counts in channel B to counts in channel A, i.e., B/A.

If the quench range to be covered is large, it is necessary to choose different discriminator settings, because moderate quench with the channel settings shown in Fig. X-6 will lead to an SCR value of zero as all pulses will be reduced in height below the lower discriminator setting of channel B. Therefore, the lower discriminator setting of channel B should be at a much lower value when the quench range is large. Figure X-7 shows possible

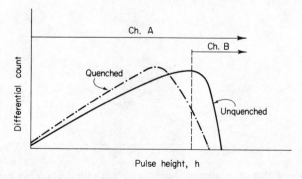

Fig. X-6. Selection of counting channels with SCR method for limited quench range.

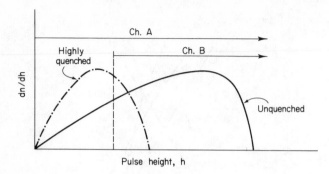

Fig. X-7.   Selection of counting channels in SCR method for wide quench range.

discriminator settings for the counting channels for samples which have a wide quench range. The change in SCR will be small for small quench differences near an unquenched sample. But considerable quench can be present without fear of the SCR value reaching a nonreal value, i.e., zero.

Plots of the SCR values for the two settings as a function of the counting efficiency are shown in Fig. X-8. Curve (a) shows a large change in SCR for small change in counting efficiency and is a sensitive indicator. However, the SCR becomes zero, while the efficiency is still quite high. Curve (b), on the other hand, is still real (nonzero SCR) after an almost tenfold decrease in the counting efficiency. At high SCR, there is little change for a large change in counting efficiency, and therefore curve (b) is not a sensitive monitor for small quench changes near unquenched level. At low counting efficiency, the change in SCR is greater and serves as a sensitive monitor for efficiency changes.

Some limits to this method are the count rate and the quench range. If the count rate is low, then the time required to obtain a statistically

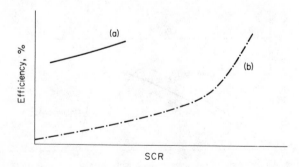

Fig. X-8.   Typical plots of counting efficiency versus SCR with counting channels as selected in (a) Fig. X-6 and (b) Fig. X-7.

reliable SCR will be very long, even longer than required to determine the activity of the sample. The statistical accuracy of the sample activity is dependent on the total counts, whereas statistical accuracy of the ratio is dependent on the reliability of the least accurately known number which is used to calculate the ratio.

External Gamma-Ray Sources

The most commonly used quench monitoring methods involve the use of external γ-ray sources which irradiate the solution. A fraction of the γ rays will undergo Compton scattering, which results in the production of a new γ ray of less energy than the original γ ray and an electron which will have an energy nearly equal to that lost by the γ ray. One reason this method is so widely used is partly because almost all commercial instruments either have external γ-ray sources or can be readily changed to include one.

The Compton scattering process will produce a continuum of electrons (somewhat like a beta continuum) with electrons of energy from zero to a maximum E_{max}, which is only dependent on the energy of the original γ ray E_{γ_0}:

$$E_{max} = \frac{2 E_{\gamma_0}^2}{2 E_{\gamma_0} + 0.51},$$

where all energies are expressed in million electron volts and 0.51 is the rest mass energy equivalent of an electron. Several different γ-ray sources are used commercially, none of which has an advantage over the others. They all interact in a like manner. The only difference is in the E_{max} of the Compton spectrum, which will differ with the E_{γ_0} of the γ-ray source. Several γ-ray sources are listed in Table X-2 along with the E_{max} of the Compton spectra.

The spectrum of pulses produced in a liquid scintillation counter by the Compton scattered electrons resulting from the interaction of the 0.662-MeV γ rays from ^{137m}Ba is shown in Fig. X-9. The pulse height distribution

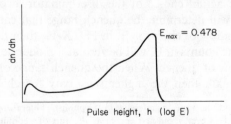

Fig. X-9. Pulse distribution for Compton scattered electrons produced by the 662-keV γ rays from ^{137m}Ba.

Table X-2

Compton E_{max} of several gamma emitters used as external standards

Gamma-ray source	Half-life	E_{γ_0} (MeV)	Percent of emitted gamma rays with the energy E_{γ_0}	Compton E_{max} (MeV)
^{137}Cs–^{137m}Ba	30 years			
	2.5 minutes	0.662	100	0.478
^{133}Ba	7.2 years	0.300	31	0.162
		0.357	69	0.209
^{22}Na	2.6 years	0.51	67[a]	0.340
		1.276	33	1.065
^{226}Ra	1622 years	Many gammas with energies from 2.4 to 0.2 when in equilibrium with daughters	Complicated to determine	Several spectra up to 2.17

[a] Two gammas of 0.51 MeV per annihilation of positron. External source geometry gives small probability of both quanta interacting with the scintillator.

is altered by quenching in a manner identical to that for a beta emitter. Increasing quench will reduce the photon yield per energy with the resulting shift in the distribution to lower pulse heights. Figure X-10 shows the distributions for the Compton scattered electrons (^{137m}Ba) with different amounts of quench. The total counts have been normalized.

There are two methods of quench monitoring with external γ-ray sources: (1) measuring the counts in a specific pulse height range (window) (8–11), and (2) measuring the ratio of counts in two specific windows.* The first is commonly referred to as the external standard counts (ESC) method and the second as the external standard channels ratio (ESCR) method.

The ESC method depends on the decrease in the counts measured in a unit time as the quench increases. This can readily be seen from the curves in Fig. X-11. The actual choice of the discriminator levels to define the counting window will determine the quench range that can be monitored. If the counting window of choice is B_1 in Fig. X-11, then the quench range will be small, as the counts in B_1 will be zero at a moderate quench and for any greater amount of quench. When the quench range of the samples to be measured is small, then the higher sensitivity for changes in quench

* There does not seem to be a specific reference in the published literature about the development of this technique. However, it did first appear as a feature of a commercial instrument in 1965 on the Beckman Instruments, Inc. LS-200, followed shortly by incorporation in the Nuclear-Chicago Mark-I.

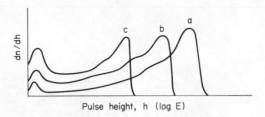

Fig. X-10. Pulse spectra for ^{137m}Ba γ rays as function of the quench level of the scintillator solution. (a) Unquenched, (b) moderate quench, (c) strong quench.

would lead to the choice of B_1. If the quench range is large, then the choice of B_2 will allow for measurements over the entire range. The relative counts will not change as much per unit of quench as for B_1.

A series of samples with known amounts of the radioactive nuclide and varying amounts of quench are measured to obtain the sample counting efficiency and the external standard count (ESC) or counts per minute (ESCPM). A typical curve for ^{3}H efficiency versus ESCPM is shown in Fig. X-12. If part of the sample counts fall within the counting window for the external standard, a correction for those counts will have to be made. Samples with unknown amounts of quench are counted with and without the external standard. The ESCPM value will project to a point on the curve which will correspond to the counting efficiency of the sample.

There are several serious limitations to the ESC method which have almost eliminated the use of it as a method of quench correction. Several of these are summarized in Table X-3.

The external standard channels ratio method (ESCR) has overcome these limitations. Since a ratio is obtained, the effects on the measured counts will be the same for both channels and hence will cancel out. The selection of the channels for counting the external standard will depend somewhat on the desired quench range and sensitivity. Figure X-13 shows at least two channel choices. There are, of course, many other combinations,

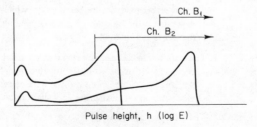

Fig. X-11. Choice of quench monitoring counting channel (B_1 or B_2) for external standard count method. B_1 for slight quench; B_2 for strong quench.

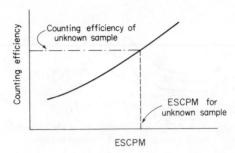

Fig. X-12. Plot of external standard count rate ESCPM versus counting efficiency. Method of obtaining counting efficiency for sample with unknown amount of quench from ESCPM of the sample.

which are equally useful. Again, it should be remembered that, if the sample introduces counts into either or both of the channels for counting the external γ-ray source, a correction has to be applied before the ratio is calculated. In some of the commercial instruments the sample contribution is automatically subtracted and the ratio calculated from the corrected counts.

Again, a counting efficiency versus ESCR plot is obtained with the aid of a series of quenched samples, with known amounts of the radioactive nuclide. Figure X-14 shows typical curves from samples containing ^3H and ^{14}C with windows as shown in Fig. X-13a. Any sample of unknown activity

Table X-3

Limitations of the ESCPM method for quench correction

Factor	Effect
Position of gamma source relative to sample	The source position has to be exactly reproduced or changes in the counts will be observed which will not reflect quench changes
Volume of sample	The measured count varies as the volume of the sample changes, because the number of γ rays, which are scattered, depends on the amount of material available with which the γ rays can interact
Half-life of the γ-ray source	The count rate has to be expressed on a relative basis or corrected for decay from the time elapsed between when the quench curve and samples were measured. This is more critical for measuring very slight quench changes
Changes in electron density of the sample and/or surroundings	The number of scattered electrons will be different if the number of electrons (electron density) in the γ-ray flux changes. The sample itself can alter the electron density, as can the thickness of the vial walls, possible quenches, etc.

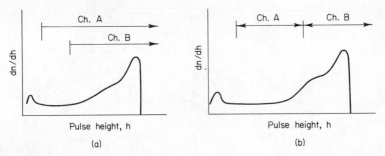

Fig. X-13. Possibilities for selection of external standard counting channels for the ratio technique.

and unknown quench can be compared to this plot. The measured ESCR will correspond to a point on the plot and thus give the counting efficiency of the ^3H and ^{14}C in the sample. Hence, the actual amount of ^3H and ^{14}C (dpm) can be calculated.

A final method of quench monitoring which uses an external γ-ray source involves the actual measurement of the pulse height corresponding to the Compton edge of the scattered electron distribution (12–17). This technique is very sensitive and accurate, but requires some method of measuring the pulse height distribution. This can be accomplished with a multichannel analyzer or by scanning with a narrow window of the liquid scintillation discriminator system. Unfortunately, not all commercial liquid scintillation counters are designed so that a multichannel analyzer can be easily employed. Such requirements as pulse shape, rise time, linearity, and other factors have to be made compatible with the input requirements of the multichannel analyzer. These requirements are not the same for all multichannel analyzers.

Scanning the pulse height distribution with a narrow window can be very time consuming if many samples are to be counted. This could increase the time per sample to an undesirable length. Also, not all commercial

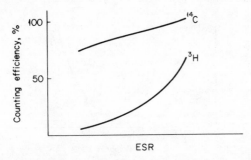

Fig. X-14. Counting efficiencies for ^{14}C and ^3H in quenched samples as a function of external standard channels ratio.

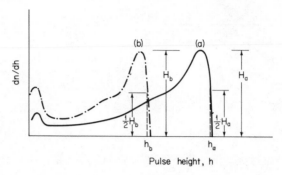

Pulse height, h

Fig. X-15. Compton pulse distributions for (a) unquenched and (b) quenched samples, showing the half-height method of obtaining Compton edge pulse height-quench monitor.

instruments have variable discriminator settings, which are necessary for such scans.

However, if it is possible to measure the pulse height corresponding to the Compton edge for a series of quenched samples with known activity, it is possible to construct a counting efficiency curve. The Compton edge for the γ rays from ^{137m}Ba is shown in Fig. X-15. The pulse height can be defined in several ways, but a convenient and reproducible method involves the method called half-height value (14). The count rate at the peak is H_a and the Compton edge is selected as corresponding to the value of $H_a/2$. The pulse height corresponding to $H_a/2$ is h_a. The value of h will vary as the quench changes. For a higher quenched sample the Compton edge might have a value h_b. This method is independent of the counting time. Since only the pulse height is measured, it is not necessary to count the

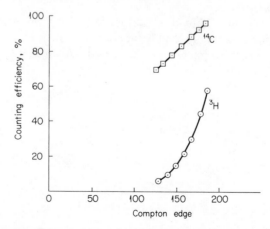

Fig. X-16. Counting efficiency of ^{3}H and ^{14}C as a function of quench as monitored by the Compton edge pulse height.

external standard for the same time for different samples. The only requirements are that enough counts be accumulated to give a statistically valid Compton distribution. It is not necessary to subtract the sample activity, as the beta continuum will not alter the pulse height of the Compton edge. One exception which might occur involves those cases where the sample count rate is very high and, when added to the count rate of the external source, might exceed the resolving time of the counter and lead to distortion due to pile-up events.

Figure X-16 shows the counting efficiency of ^3H and ^{14}C as a function of the Compton edge pulse height. From such plots the counting efficiency of any sample can be obtained by measuring the pulse height of the Compton edge on that sample.

One modification of this method involves the use of the Compton edge pulse height and integral counting (14). An unquenched ^{14}C-containing sample was counted by the integral counting technique (18) and the pulse height of the Compton edge measured as $(PH)_0$. The sample was quenched (with ethanol), recounted, and the new pulse height of the Compton edge measured $(PH)_1$. This was repeated for several increasing amounts of quench. The results were then plotted in Fig. X-17 with the counting efficiency as a function of the ratio of $(PH)_0/(PH)_n$. The relationship was linear over a sevenfold quench. The counting efficiency of any unknown sample could be obtained from this plot and a measure of $(PH)_n$.

Other Methods

Several other methods which are also used to monitor the amount of quench in a counting sample are efficiency stick (19, 20), optical filter (21),

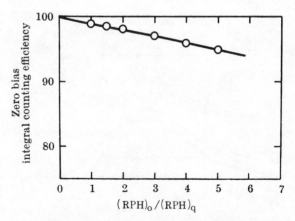

Fig. X-17. Integral counting rates of unquenched and ethanol-quenched ^{14}C samples (equal ^{14}C concentration) as a function of relative pulse heights of Compton edge.

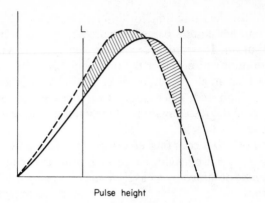

Pulse height

Fig. X-18. Balance point counting showing selection of counting channel (L to U) such that the two shaded areas are equal. Thus counts lost from counting channel upon quenching (dashed curve) equal counts gained. This will provide constant counting efficiency.

optical absorbency (22–26), tracer (27–29), double channels ratio (30), and extrapolation (31–34).

Techniques of Constant Counting Efficiency

Some experimental counting techniques have been used for which the counting efficiency remains nearly constant over a given quench range. Three of these techniques are discussed here.

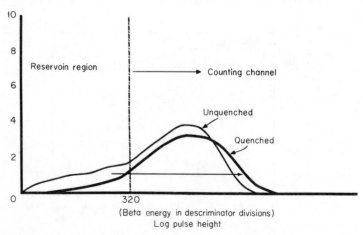

Fig. X-19. Overrestoration of quenched ^{14}C sample to give constant efficiency in counting channel (39).

Table X-4

Counting efficiency of quenched ^{14}C samples using gain overrestoration to give constant efficiency[a]

Added CHCl$_3$ (ml)	Efficiency (%)	
	Without gain restoration	With gain overrestoration
0	60	60
0.05	51	61
0.1	42	60
0.2	35	61
0.3	27	59
0.4	21	59
0.5	16	58

[a] See C. H. Wang, *in* "The Current Status of Liquid Scintillation Counting (E. D. Bransome, Jr., ed.), p. 305. Grune & Stratton, New York, 1970.

Balance Point Counting[*]

Selection of the counting channel as shown in Fig. X-18 has been shown to lead to constant counting efficiency over a limited quench range. This method is based on the fact that under certain conditions the quench will lead to a constant number of counts in the window (L to U) because the counts lost at the lower discriminator L are exactly equal to the counts gained at the upper discriminator U.

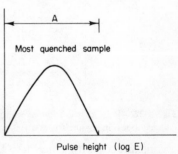

Fig. X-20. Setting counting channel A for total counts of most quenched samples.

This technique has proved to be very useful for samples with small differences in quench which might be difficult to determine by small changes in sample channels ratio or external standard channels ratio. However, the count rate of the samples has to be high because only a fraction of the sample counts are actually measured. Therefore, longer counting times are required to reach the desired counting statistics. For low count rate samples this often proves to be a serious drawback.

[*] See Arnold (35) and Packard (36).

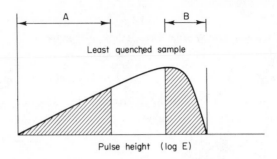

Pulse height   (log E)

Fig. X-21.   Setting counting channel B for sum of counts (shaded areas) in channels A and B equal to counts of most quenched sample in channel A.

Gain Overrestoration*

The technique of changing the gain of the detection system of a liquid scintillation counter is usually used to reestablish the relationship of pulse height (energy) and the discriminator settings of the counting channels. However, if the gain is overrestored, it is possible over a limited quench range to obtain a constant counting efficiency. A measure of some quench monitor, such as sample channels ratio or external standard channels ratio, is used to change the gain of the detector system automatically to restore

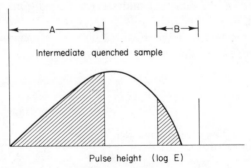

Pulse height   (log E)

Fig. X-22.   Pulse distribution of sample with intermediate quench whereas sum of counts (shaded areas) in channels A and B is equal to counts of most quenched sample in channel A.

a constant counting efficiency. The gain change is predetermined by a series of standard quenched samples. Table X-4 shows the results obtained by Wang (39), using the technique of gain overrestoration for counting ^{14}C samples quenched with $CHCl_3$. The counting channel used by Wang is shown in Fig. X-19.

This technique cannot be utilized with the low-energy beta emitter ^{3}H, because no amount of gain increase will reestablish counts for beta particles that do not produce enough photons to give at least a single photoelectron

* See the literature (37–40).

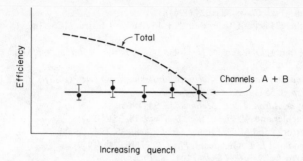

Fig. X-23. Constant counting efficiency by the two-channel method. Total efficiency shows the sacrificing of some counts to obtain constant efficiency.

in each MPT. However, the technique is quite applicable to higher-energy beta emitters, over a limited quench range.

Double Counting Channels

The use of two counting channels to give constant counting efficiency has been proposed by Rapkin (41). However, in this method the counting efficiency can be no greater than that of the most quenched sample. One counting channel is set by discriminators which bracket the total pulse height distribution of the most quenched sample. Figure X-20 shows the settings of the discriminators for channel A. The second counting channel is set using the unquenched (or least quenched) sample. Figure X-21 shows the pulse distribution of the unquenched sample relative to the counting channel A. The upper and lower levels of the counting channel B are set as shown in Fig. X-21 so that the total counts in channels A and B for the unquenched sample are equal to the counts of the quenched sample in channel A only. Finally Fig. X-22 shows the pulse distribution in the two channels for a sample with intermediate quench.

Once the channels are selected, a series of samples of known dpm and varying quench are counted and the total counts in channels A and B recorded. Figure X-23 shows some results reported by Rapkin. Of course, by sacrificing total counts, the counting time required to obtain a certain required statistical reliability will be increased.

References

1. F. N. Hayes, *Int. J. Appl. Radiat. Isotop.* **1,** 46 (1956).
2. G. T. Okita, J. Spratt, and G. V. Leroy, *Nucleonics* **14,** No. 3, 76 (1956).
3. J. D. Davidson and P. Feigelson, *Int. J. Appl. Radiat. Isotop.* **2,** 1 (1957).
4. M. L. Whisman, B. H. Eccleston, and F. E. Armstrong, *Anal. Chem.* **32,** 484 (1960).
5. L. A. Baillie, *Int. J. Appl. Radiat. Isotop.* **8,** 1 (1960).

6. E. T. Bush, *Anal. Chem.* **35,** 1024 (1963).
7. J. R. Herberg, Packard Tech. Bull. No. 15. Packard Instrum. Co., Inc., Downers Grove, Illinois, 1965.
8. T. Higashimura, O. Yamada, N. Nohara, and T. Shidei, *Int. J. Appl. Radiat. Isotop.* **13,** 308 (1962).
9. D. G. Fleisman and U. V. Glazunov, *Instrum. Exp. Tech. (USSR)* p. 472 (1962); *Prib. Tekh. Eksp.* **3,** 55 (1962).
10. F. N. Hayes, *Advan. Tracer Methodol.* **3,** 95 (1966).
11. R. De Wachter and W. Fiers, *Anal. Biochem.* **18,** 351 (1967).
12. D. L. Horrocks and M. H. Studier, *Anal. Chem.* **30,** 1747 (1958).
13. K. F. Flynn and L. E. Glendenin, *Phys. Rev.* **116,** 744 (1959).
14. D. L. Horrocks, *Nature (London)* **202,** 78 (1964).
15. D. L. Horrocks, *Nucl. Instrum. Methods* **27,** 253 (1964).
16. D. L. Horrocks, *Nucl. Instrum. Methods* **30,** 157 (1964).
17. K. F. Flynn, L. E. Glendenin, E. P. Steinberg, and P. M. Wright, *Nucl. Instrum. Methods* **27,** 13 (1964).
18. J. Steyn, *Proc. Phys. Soc. London Sect. A* **69,** 865 (1956).
19. W. J. Kaufman, A. Nir, G. Parks, and R. M. Hours, *Proc. Conf. Organic Scintillation Detectors, Univ. of New Mexico, 1960* (G. H. Daub, F. N. Hayes, and E. Sullivan, eds.), TID-7612, p. 239. U. S. At. Energy Commission, Washington, D. C., 1961.
20. H. E. Dobbs, *Nature (London)* **200,** 1283 (1963).
21. K. F. Flynn, L. E. Glendenin, and V. Prodi, *in* "Organic Scintillators and Liquid Scintillation Counting" (D. L. Horrocks and C. T. Peng, eds.), p. 687. Academic Press, New York, 1971.
22. S. Helf and C. White, *Anal. Chem.* **29,** 13 (1957).
23. R. J. Herberg, *Anal. Chem.* **32,** 1468 (1960).
24. H. H. Ross and R. E. Yerick, *Anal. Chem.* **35,** 794 (1963).
25. J. De Bersaques, *Int. J. Appl. Radiat. Isotop.* **14,** 173 (1963).
26. H. H. Ross, *Anal. Chem.* **37,** 621 (1965).
27. O. Yura, *Radioisotopes* **20,** 383 (1971).
28. O. Yura, *Radioisotopes* **20,** 493 (1971).
29. O. Yura, *Radioisotopes* **20,** 610 (1971).
30. E. T. Bush, *Int. J. Appl. Radiat. Isotop.* **19,** 447 (1968).
31. C. T. Peng, *Anal. Chem.* **32,** 1292 (1960).
32. C. T. Peng, *Anal. Chem.* **36,** 2456 (1964).
33. C. T. Peng, *in* "Organic Scintillators" (D. L. Horrocks, ed.), p. 109. Gordon & Breach, New York, 1968.
34. M. A. Dugan and R. D. Ice, *in* "Organic Scintillators and Liquid Scintillation Counting" (D. L. Horrocks and C. T. Peng, eds.), p. 1055. Academic Press, New York, 1971.
35. J. R. Arnold, *Science* **119,** 155 (1954).
36. L. E. Packard, *in* "Liquid Scintillation Counting" (C. G. Bell, Jr. and F. N. Hayes, eds.), p. 50. Pergamon, Oxford, 1958.
37. C. H. Wang, *Advan. Tracer Methodol.* **1,** 285 (1962).
38. C. H. Wang, Atomlight, No. 21. New England Nucl. Corp., Boston, Massachusetts, 1962.
39. C. H. Wang, *in* "The Current Status of Liquid Scintillation Counting" (E. D. Bransome, Jr., ed.), p. 305. Grune & Stratton, New York, 1970.
40. P. Jordon, U. Kaczmar, and P. Köberle, *Nucl. Instrum. Methods* **60,** 77 (1968).
41. E. Rapkin, Private communication, 1968.

CHAPTER XI

DUAL-LABELED COUNTING

Often it is desirable to use two different radionuclides in the same experiment to follow different functions of the same system at the same time. And as a result, there is the problem of counting the two activities together or separating the activities for counting each one individually. Often the two radionuclides are different elements (i.e., ^{14}C and ^{3}H, ^{32}P and ^{35}S, etc.). However, it is just as likely that the two radionuclides are isotopes of the same element (i.e., ^{241}Pu and ^{239}Pu, ^{33}P and ^{32}P, etc.).

Simultaneous Determination

In order to determine each of the two radioactivities in the same sample, there has to be a difference in some nuclear property of the two radionuclides. Some examples are given in Table XI-1.

Table XI-1

Examples of radionuclide pairs that can be determined simultaneously in a liquid scintillator system because of different nuclear properties

Dual label	Different nuclear properties
^{241}Pu–^{239}Pu	^{241}Pu: 0.021-MeV E_{max} β^- decay
	^{239}Pu: 5.1-MeV alpha decay
^{3}H–^{14}C	^{3}H: 0.018-MeV E_{max} β^- decay
	^{14}C: 0.158-MeV E_{max} β^- decay
^{125}I–^{131}I	^{125}I: 60-day half-life, low-energy electrons
	^{131}I: 8-day half-life, 0.25–0.81-MeV E_{max} β^- decay
^{33}P–^{32}P	^{33}P: 25-day half-life and 0.251-MeV E_{max} β^- decay
	^{32}P: 14-day half-life and 1.707-MeV E_{max} β^- decay

Different Particle Emission

Horrocks and Studier (1) showed an early example of dual-labeled counting in the determination of the isotopes ^{241}Pu and ^{239,240}Pu. The ^{239,240}Pu isotopes decay by the emission of 5.1-MeV alpha particles while the ^{241}Pu decays by β^- emission with an E_{max} of 0.021 MeV. Figure XI-1 shows the counting results from a sample of plutonium dissolved in a liquid scintillator solution. The low-energy beta particles of ^{241}Pu produced a continuum of pulses of small amplitude (similar to ^{3}H), whereas the 5.1-MeV alpha particles produced a peaked distribution of pulses of large amplitude. The two spectra were easily isolated into two counting channels for measurement of the two radioactivities.

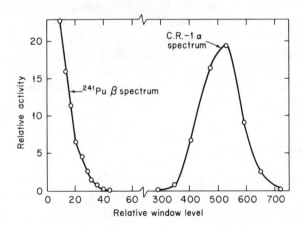

Fig. XI-1. Differential pulse height distribution for a plutonium sample (containing the isotopes 239–241) in 0.25 ml of a liquid scintillator solution.

Different Half-Life

It is difficult to measure ^{125}I and ^{131}I together in a liquid scintillator solution because of the complex decay scheme of each nuclide and the overlap of the ^{131}I beta decay spectrum with the conversion and Auger electrons emission spectrum of ^{125}I. One alternative method of measuring both ^{125}I and ^{131}I in the same sample depends on the difference in half-lives, which are 60 days and 8 days, respectively. The sample can be counted at two different times and the use of simultaneous equations will relate the difference in count rate to the activity of each nuclide. If C_0 is the measured count rate of the sample (^{125}I and ^{131}I) at the starting time, and C_1 the measured count rate at a later time, then

$$C_0 = \epsilon_1 A_{10} + \epsilon_2 A_{20}, \quad \text{and} \quad C_1 = \epsilon_1 A_{11} + \epsilon_2 A_{21},$$

where ϵ_1 is the counting efficiency of isotope 1, ϵ_2 the counting efficiency of isotope 2, A_{10} the activity (dpm) of isotope 1 at time 0, A_{11} the activity (dpm) of isotope 1 at time 1, A_{20} the activity (dpm) of isotope 2 at time 0, A_{21} the activity (dpm) of isotope 2 at time 1, and

$$A_{11} = A_{10} \exp(-\lambda_1 t), \quad A_{21} = A_{20} \exp(-\lambda_2 t),$$

where t is the time elapsed between measurement 0 and 1, and λ is the decay constant of the nuclide

$$\lambda = 0.693/t_{1/2}$$

or

$$\lambda(^{125}\text{I}) = 0.693/60 \text{ days}, \quad \lambda(^{131}\text{I}) = 0.693/8 \text{ days}.$$

Example 1. A sample containing ^{125}I and ^{131}I was counted twice with 8 days decay between the two measurements:

$$C_0 = 6000 \text{ cpm}, \quad C_1 = 3412 \text{ cpm}.$$

What are the initial activities of ^{125}I and ^{131}I?

$$C_0 = 6000 = \epsilon_1 A_{10} + \epsilon_2 A_{20}$$

$$C_1 = 3412 = \epsilon_1 A_{11} + \epsilon_2 A_{21} = \epsilon_1 A_{10} \exp[-\lambda_1(8)] + \epsilon_2 A_{20} \exp[-\lambda_2(8)]$$

$$= \epsilon_1 A_{10} \exp\left[\frac{-(0.693)(8)}{60}\right] + \epsilon_2 A_{20} \exp\left[\frac{-(0.693)(8)}{8}\right]$$

$$= 0.912\epsilon_1 A_{10} + 0.500\epsilon_2 A_{20}.$$

Solving C_0 for A_{10} gives

$$A_{10} = (6000 - \epsilon_2 A_{20})/\epsilon_1.$$

Substituting this into the equation for C_1 gives

$$3412 = \frac{0.912\epsilon_1}{\epsilon_1}(6000 - \epsilon_2 A_{20}) + 0.500\epsilon_2 A_{20}$$

$$= 5472 - 0.912\epsilon_2 A_{20} + 0.500\epsilon_2 A_{20}$$

$$= 5472 - 0.412\epsilon_2 A_{20}$$

or

$$A_{20} = \frac{(5472 - 3412)}{0.412\epsilon_2} = \frac{2060}{0.412\epsilon_2} = \frac{5000}{\epsilon_2}$$

and

$$A_{10} = \frac{6000 - \epsilon_2(5000/\epsilon_2)}{\epsilon_1} = \frac{1000}{\epsilon_1}.$$

Thus at time 0 there were

$$
\begin{array}{r}
1000 \text{ cpm of } ^{125}\text{I} \\
+ 5000 \text{ cpm of } ^{131}\text{I} \\
\hline
6000 \text{ cpm}
\end{array}
$$

and 8 days later there were

$$
\begin{array}{r}
912 \text{ cpm of } ^{125}\text{I} \\
+ 2500 \text{ cpm of } ^{131}\text{I} \\
\hline
3412 \text{ cpm.}
\end{array}
$$

Example 2. A sample containing ^{3}H and ^{125}I was counted on different days.

$$C_0 = 7000 \text{ cpm}$$

$$C_1 = 6942 \text{ cpm} \qquad (1 \text{ day later})$$

$$C_5 = 6721 \text{ cpm} \qquad (5 \text{ days later}).$$

What was the initial count rate of ^{3}H and ^{125}I?

In this case $e^{-\lambda t}$ will be 1.00 for ^{3}H since even 5 days is insignificant compared to the 12.24-year half-life of ^{3}H. Thus the counts of ^{3}H will remain constant. So,

$$C_0 = C_T + \epsilon_2 A_{20} = 7000$$

$$C_1 = C_T + \epsilon_2 A_{21} = 6942$$

$$C_5 = C_T + \epsilon_2 A_{25} = 6721$$

and

$$A_{21} = A_{20}e^{-\lambda(1)} = 0.9884 A_{20}, \qquad A_{25} = A_{20}e^{-\lambda(5)} = 0.9442 A_{20}.$$

Thus

$$7000 - \epsilon_2 A_{20} = 6942 - 0.9884\epsilon_2 A_{20} = 6721 - 0.9442\epsilon_2 A_{20},$$

which gives

$$0.0116\epsilon_2 A_{20} = 58 \quad \text{and} \quad 0.0558\epsilon_2 A_{20} = 279.$$

Thus

$$\epsilon_2 A_{20} = 58/0.0116 = 5000$$

or

$$\epsilon_2 A_{20} = 279/0.0558 = 5000.$$

The ^{3}H activity produced 2000 cpm, and the ^{125}I activity produced 5000 cpm. The activity of each is obtained by dividing each cpm by the appropriate counting efficiency which is obtained with identical samples containing a known amount of each isotope.

Different Counting Methods

Brown (2) gave an example of a dual-labeled sample which utilizes two measurement techniques to obtain the activity of ^{33}P and ^{32}P. The ^{32}P was measured by Cerenkov counting (see Chapter XIV) of the high-energy β^- emissions, $1.707 E_{max}$. Since the threshold for production of Cerenkov radiation in water is 0.265 MeV and the maximum β^- particles from the decay of ^{33}P are 0.250 MeV, no Cerenkov radiation was produced by ^{33}P decay in a water solution (limit of $\leq 0.01\%$). Once the ^{32}P activity is known, the dual channel counting of the sample in LS counts will give the count rate of ^{33}P and ^{32}P in the window designated channel A in Fig. XI-2. Previous calibration of ^{33}P and ^{32}P efficiency in channel A will make it possible to correct for counts ^{32}P contributed to channel A. The remaining cpm

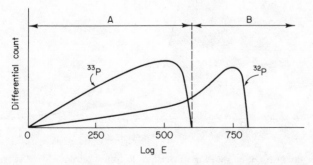

Fig. XI-2. Selection of counting channel A for counting ^{33}P in liquid scintillator solution. The spillover of ^{32}P is corrected from known dpm obtained from Cerenkov counting.

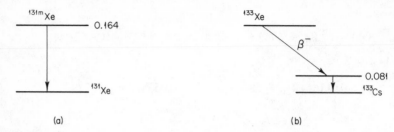

(a)   (b)

Fig. XI-3.   Decay schemes for ^{131m}Xe and ^{133}Xe. (a) ^{131m}Xe: $t_{1/2} = 12.0$ days, $e/\gamma = 40$. (b) ^{133}Xe: $t_{1/2} = 5.3$ days, E_{max} of $\beta^- = 0.347$ MeV, e/γ (0.081) = 1.8.

will be due to ^{33}P decays. The activity of ^{33}P can be calculated from the cpm and predetermined efficiency. [*Note.* The same result could be obtained by a single dual-channel counting in the LS system. This example is used to demonstrate a technique.]

Different Modes of Decay

Horrocks and Studier (3) studied the determination of the two isotopes of xenon, ^{133}Xe and ^{131m}Xe, which both decay by the emission of electrons. ^{133}Xe has a normal β^- emission, whereas ^{131m}Xe emits electrons through the internal conversion process. Figure XI-3 details the decay properties of each nuclide. The method of determining each Xe radionuclide in the presence of the other is based on the different distribution of pulses from

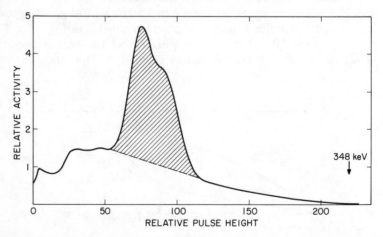

Fig. XI-4.   Pulse distribution showing conversion electron spectrum of ^{131m}Xe (shaded area) and the beta continuum of ^{133}Xe.

the LS system for monoenergetic electrons (from decay of ^{131m}Xe) and from the continuum of electrons (from decay of ^{133}Xe). Figure XI-4 shows a pulse height distribution from a sample of Xe containing ^{131m}Xe and ^{133}Xe. The pulses resulting from the conversion electrons give a peaked distribution indicated by the shaded area. The beta decay gives rise to a continuum of pulses which can be predicted from the beta spectrum. Thus the ^{131m}Xe conversion electron decays are equal to the shaded area, since the conversion electrons are counted with 100% efficiency. The ^{131m}Xe disintegration rate is calculated from the known ratio of conversion electron emissions. The ^{133}Xe count rate is equal to the area under the rest of the distribution. The ^{133}Xe disintegration rate is obtained from the previously determined counting efficiency of a sample with a known count of only ^{133}Xe.

¹⁴C–³H Dual Label

By far the most common type of dual isotope counting is that involving the determination of ^{3}H and ^{14}C in the same sample. Since both of these isotopes decay by emission of a beta continuum (^{14}C having an E_{max} 8.5 times greater than that for ^{3}H), there will be only a part of the pulse height spectrum for which the pulses will be produced by both ^{14}C and ^{3}H decays. This part of the pulse spectrum will be from zero to that corresponding to the maximum for the highest-energy beta particles from ^{3}H, 18 keV. Figure XI-5 shows a pulse height distribution from samples containing only ^{3}H, only ^{14}C, and ^{3}H and ^{14}C together. It is evident from this figure that part of the ^{14}C spectrum can be counted without any contribution to the observed counts from the decay of ^{3}H. However, no part of the ^{3}H spectrum can be selected which will have no counts due to the decay of ^{14}C.

The technique most commonly used for dual ^{14}C–^{3}H-labeled samples involves setting two channels for counting. In the upper channel (set with its lower discriminator above the ^{3}H maximum pulse height) only ^{14}C pulses are counted. In the lower window (set with its upper discriminator near the ^{3}H maximum pulse height) pulses due to ^{3}H and ^{14}C decays are counted. Figure XI-5 shows one setting of the channels. First, samples as nearly identical as possible to the unknown samples are prepared which contain known amounts of only ^{14}C in one vial (or set of vials) and only ^{3}H in another vial (or set of vials). The ^{14}C sample will be used to determine the counting efficiency from ^{14}C in the two channels. The ^{3}H sample will be used to determine the counting efficiency for ^{3}H in channel A. When the unknown sample is counted (assuming the quench and composition are identical with the standard), the counts in channel B are used to calculate

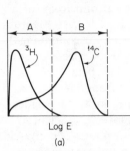 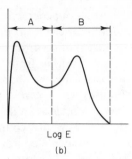

Fig. XI-5.   Relative pulse distributions for single-labeled ^{3}H and single-labeled ^{14}C containing sample (a) compared with sample labeled with both ^{3}H and ^{14}C (b).

the ^{14}C dpm and the ^{14}C counts which occur in channel A. The difference between the observed counts in channel A and those due to ^{14}C will be the number of ^{3}H counts in channel A. This, used with the ^{3}H efficiency in channel A, will give the dpm of ^{3}H in the sample. In this manner the dpm of both ^{3}H and ^{14}C can be determined in the same sample.

Example 3.   Standard solutions of ^{14}C and ^{3}H were measured in previously set channels. The efficiencies calculated were

$$\text{channel B} - {}^{14}\text{C Eff} = 75\%$$

$$\text{channel A} - {}^{14}\text{C Eff} = 23\%$$

$$\text{channel A} - {}^{3}\text{H Eff} = 61\%.$$

An unknown sample with ^{3}H and ^{14}C of identical composition was counted, giving

$$\text{channel B} = 10,000 \text{ cpm}, \qquad \text{channel A} = 7000 \text{ cpm}.$$

What were the dpm values for ^{3}H and ^{14}C?

$$^{14}\text{C dpm} = 10,000/0.75 = 13,333 \text{ dpm}$$

$$^{14}\text{C cpm in channel A} = (13,333)(0.23) = 3070 \text{ cpm}$$

$$^{3}\text{H cpm in channel A} = 7000 - 3070 = 3930 \text{ cpm}$$

$$^{3}\text{H dpm} = 3930/0.61 = 6440 \text{ dpm}.$$

The amount of ^{3}H in the sample was 6440 dpm, and the amount of ^{14}C was 13,333 dpm.

One major difficulty in the determination of dual-labeled activities is the problem of quenching. Quenching not only decreases the counting efficiency

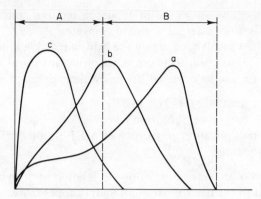

Fig. XI-6. Pulse spectra for ^{14}C-containing samples (a) unquenched, (b) intermediate quench, and (c) highly quenched showing the shift in measured counts from channel B ($^{14}C/^3H$) to channel A (^3H) with increasing quench.

of a radionuclide like ^{14}C or ^3H, but it leads to a change in the distribution of pulse height for the measured scintillations. Figure XI-6 shows the shift in the differential pulse height spectrum for ^{14}C-containing samples with different amounts of quenching. If the windows defining the counting channel, such as channel B, are set as shown in Figs. XI-5 and XI-6, there will actually be an increasing fraction of the total counts lost from the measured counts in channel B.

There are two commonly used techniques which allow for the measurement of the maximum number of counts (or pulses). The first is a change of the window settings as the quench changes. Figure XI-7 shows a pair of window settings for counting two samples of different quench. The window settings can be predetermined with a series standard quenched sample

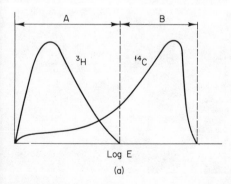

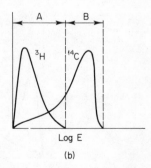

Fig. XI-7. Window settings for counting ^3H- and ^{14}C-containing samples with optimization of separation of ^{14}C- and ^3H-produced pulses, showing settings for (a) unquenched sample and (b) quenched sample.

correlating the window settings with a quench monitor, such as external standard channels ratio, external standard cpm, etc.

A second method involves changing the gain of the system to reestablish a predetermined relationship between a set of windows and the counting efficiency. The gain can be adjusted by one of three different means:

1. change in high voltage (HV) on MPT,
2. change in electronic amplification gain,
3. change in multiplication properties of MPT by other than HV, i.e., use magnetic defocusing.

The restoration of the pulse distribution in relation to the window setting is very important to the determination of dual isotopes in a single counting sample. Figure XI-8 shows the effect of quenching on the pulse distributions of ^{3}H- and ^{14}C-labeled sample with constant counting channels.

It will be noted that the ^{3}H counts remain in the lower channel but that the distribution is shifted to lower pulse height. Also, there are still only ^{14}C counts in the upper channel. However, the fraction of ^{14}C counts in

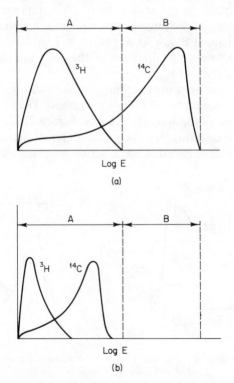

Fig. XI-8. Pulse distributions for ^{3}H and ^{14}C from (a) unquenched and (b) quenched samples relative to fixed window settings and the same system gain.

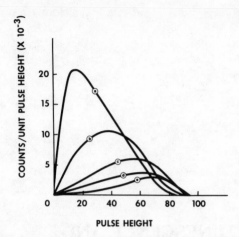

Fig. XI-9. Pulse height distribution for ³H pulses from samples of different quench level (indicated by their S values) with gain restored for a predetermined fraction of ³H dpm. (a) $S = 0.800$ (unquenched), (b) $S = 0.640$, (c) $S = 0.480$, (d) $S = 0.360$, (e) $S = 0.140$.

the upper channel is greatly reduced, which means longer counting times to reach a statistically meaningful ¹⁴C determination. The fraction of ¹⁴C counts in the lower channel is greatly increased, which means that the determinations of the number of ³H counts in the lower channel will be known with less accuracy because of the uncertainty of the difference between two large numbers.

Using one of the restoration methods, the pulse distribution relative to the window settings can be restored almost completely. Figure XI-9 shows the spectrum for a restored (e.g., change HV on MPT) distribution of ³H at different amounts of quench as indicated by the external standard channels ratio, S.

Separation of Labeling Radionuclides

If the two labeling radionuclides can be separated, then each can be determined without interference from the other. Combustion techniques have been used to separate ³H (as HTO) and ¹⁴C (as ¹⁴CO₂). Chromatographic techniques, column elutions, solubility differences, and other separation techniques can be used if the radionuclides are on different compounds.

References

1. D. L. Horrocks and M. H. Studier, *Anal. Chem.* **30,** 1747 (1958).
2. L. C. Brown, *Anal. Chem.* **43,** 1326 (1971).
3. D. L. Horrocks and M. H. Studier, *Anal. Chem.* **36,** 2077 (1964).

CHAPTER XII

CHEMILUMINESCENCE AND BIOLUMINESCENCE

Chemiluminescence and bioluminescence result from reactions which lead to a product that is in an excited state. Some of these excited molecules are able to deactivate by emission of photons; others are able to transfer their excitation energy to other molecules in the reaction system which are able to emit photons. The overall effect is the production of photons, which is directly related to the chemical reaction that occurred. The yield of photons will be affected by several factors:

1. rate of reaction (concentration of reactants, temperature, catalysts, etc.),
2. side reactions (impurity concentrations, competitive reaction rates),
3. fluorescence yield of excited species,
4. energy transfer efficiency (if primary excited species is not able to emit photons), and
5. optical transmission of photons by the reaction medium.

Basic Processes

Chemiluminescence and bioluminescence can be viewed as a process by which certain excited electronic states are produced by the energy of a chemical reaction. The excited electronic state can undergo several competitive modes of energy dissipation. Most chemical reactions convert the excess energy into kinetic energy (thermal dissipation). The luminescence-producing reactions convert the excess energy (or a fraction of it) into photons. There are very few reactions known which actually lead to the observation of photons.

Chemiluminescence

Most of the earlier known chemiluminescence reactions involved either oxygen or peroxide, and these were thought to be essential to these types of reactions. Recent studies (1) have shown that other types of reactions can also lead to luminescence. Some of these reactions involve chemicals of biological systems (usually referred to as bioluminescence).

There are different techniques used by which reactions can be made to yield photons. One is to add to the system a molecule which acts as an acceptor of energy from an excited product of the reaction which is not an efficient fluorescence molecule. The acceptor molecule, which has a high fluorescence yield, then emits the photons. A second technique is to add to the reactant molecule a species which is able to fluoresce by concentrating the excitation energy, but which does not influence the chemical reaction (metal chelates).

If the reaction is to produce photons, it is necessary that

1. sufficient energy be produced for excitation,
2. a species be present which is capable of forming an excited electronic state,
3. a species be present which is capable of radiating the energy (fluorescing),
4. the reaction rate be rapid enough to be able to measure the photons, and
5. the side or competitive reactions be less favorable than the luminescence-producing step.

Consider a simple reaction of two species, A and B, present in a system (this might be a liquid scintillator solution):

$$A + B \rightarrow C^* + D, \qquad C^* \rightarrow C + h\nu$$

where one of the products is produced in an electronically excited state C^*.

The excited molecule is able to deexcite by converting its excess energy into a photon of energy $h\nu$. The chemiluminescence yield Φ_{CL} is the ratio of the number of photons to the molecules of reactant used in the reaction:

$$\Phi_{CL} = \frac{(\text{number of photons}/N)}{\text{moles of A (or B) reacted}}$$

where N is Avogadro's number. The actual value of Φ_{CL} is given by the product of two different yields:

$$\Phi_{CL} = \Phi_{es} \cdot \Phi_f$$

where Φ_{es} is the yield of the excited states by the chemical reaction and Φ_f is the fluorescence yield of the excited state.

Chemiluminescence is best viewed as a process for the production of excited molecules. It has been shown in many chemiluminescence reactions that the excited electronic states produced by the chemical reaction are the same ones involved in normal scintillation events or optical excitation of the excited electronic levels. The process can be discussed from the consideration of a simple chemical reaction involving an electron transfer, such as an oxidation–reduction reaction (1). In normal photoexcitation it can be considered that an electron is given extra energy and promoted into one of the higher-energy electronic levels:

(a) Photoexcitation followed by fluorescence:

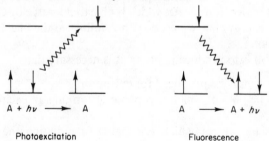

Photoexcitation   Fluorescence

(b) A chemical excitation by electron transfer (ion annihilation):

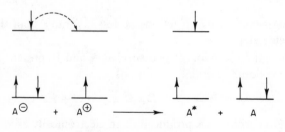

Electron transfer

followed by fluorescence:

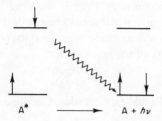

It has been shown (1) that the spectrum of photons produced in a chemiluminescence reaction is identical to that produced by photoexcitation of the first excited singlet electronic energy level. The chemiluminescence reaction was the ion annihilation of rubrene:

$$\text{Rubrene}^{\oplus} + \text{Rubrene}^{\ominus} \rightarrow \text{Rubrene}^* + \text{Rubrene}$$

$$\text{Rubrene}^* \rightarrow \text{Rubrene} + h\nu.$$

The spectrum of photon intensity versus wavelength from the ion-annihilation reaction is shown in Fig. XII-1. It is identical to the fluorescence spectrum of rubrene, also shown in the figure.

In many chemical reactions an excited species is produced which is not observed because the excitation energy goes into other processes instead of being released as photons. In many cases an energy acceptor can be added to the system which will accept the energy via a transfer process. The excited acceptor can then emit the extra energy as photons in a more efficient manner. Of course this is why the liquid scintillator solution is an efficient system for detecting chemi- and bioluminescence reactions. The presence of the very efficient energy acceptors and emitters, as the scintillator solutes, greatly enhances the probability of observing the excited molecules produced by the reactions. However, the requirements for efficient energy

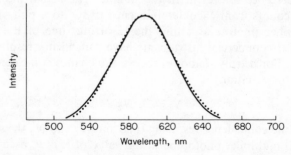

Fig. XII-1. Emission spectra for optically excited rubrene ($-$) and rubrene excited by the ion-annihilation reaction ($\cdots$).

transfer have to be fulfilled. If the excited species produced by the chemical reaction has less excitation energy than is required to produce the excited scintillator solute molecule, no fluorescence will be observed. Also the concentration of the acceptor molecules has to be sufficiently great to provide a high probability of energy transfer.

Chemiluminescence and Liquid Scintillation Counting

Chemiluminescence, for most investigators, presents a great deal of trouble for the accurate determination of the observed counts that are due only to the radioactive label of the sample. The presence of chemiluminescence is most often associated with the type of sample or the type of treatment given to the sample in its preparation for counting in the liquid scintillator solution. However, there are also those investigators who are using liquid scintillation counters to measure the concentration of certain reactants which create luminescence through chemical reactions.

There are some major misconceptions concerning the measurement of chemiluminescence in liquid scintillation counters. All of the following are *false*:

1. that the photon yield of chemiluminescence can be related to the energy of beta particles,
2. that a single count corresponds to a single photon,
3. that decreases in temperature eliminate chemiluminescence,
4. that counts due to chemiluminescence can be eliminated by operation in coincidence with a low-energy bias.

The photons produced by these chemical reactions are random in time. They can be measured in a coincidence-type liquid scintillation counter only if the photon intensity is quite high. Thus switching the instrument to count in a singles (noncoincidence) mode should give a high count rate if chemiluminescence is present in the sample. The actual measurement of coincident counts from chemiluminescence is due to a pulse produced in each multiplier phototube within the resolving time of the coincidence system. Most commercial instruments have coincidence resolving times of 20–30 nsec. For a truly random process, the chance coincidence rate (N_c) is given by the equation:

$$N_c = 2\tau_r N_2 N_2$$

where τ_r is the coincidence resolving time and N_1 and N_2 the singles count rate in each multiplier phototube. If the value of τ_r is 30 nsec and N_1 and

N_2 are 2×10^5 cpm, the chance coincidence rate will be

$$N_c = 2\left(\frac{30 \times 10^{-9}}{60} \min\right)(2 \times 10^5)(2 \times 10^5)$$

$$= 40 \text{ cpm.}$$

From this calculation it is easy to see that the singles rate is very high with small increases in the coincidence count rate. Thus if high to moderate count rate samples are being measured, the chemiluminescence rate can be high without producing a significant error in sample activity. The measurement of low count rate samples is much more critical if any chemiluminescence is present.

Consider how a coincidence pulse can be produced from a random process. The pulse produced in one multiplier phototube opens the coincidence gate for 30 nsec. If a pulse occurs in the second multiplier phototube during that 30 nsec, a coincidence count will be recorded:

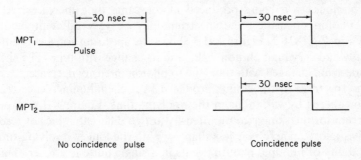

Further, the amplitude of the coincidence pulse will be a measure of the number of single events (photoelectrons) which occurred during the 30 nsec that the pulse is integrated:

With pulse summation the final coincidence pulse will be equivalent to the sum of the number of single events in each multiplier phototube within the 30 nsec after the opening of the coincidence gate by a single pulse within either of the two multiplier phototubes, MPT_1 and MPT_2:

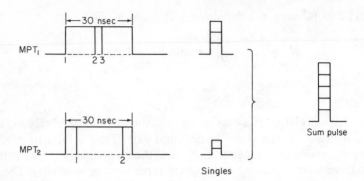

The distribution of pulse amplitudes from a single multiplier phototube will be a function of the rate of chemiluminescence. If the photon intensity per unit time is low, the preponderance of events will correspond to single photoelectron pulses. Figure XII-2 shows a typical spectrum of a solution with low to moderate chemiluminescence photon intensity.

As the intensity of the photons increases, the probability of more than one photoelectron being produced at the same time increases. At very high photon intensities, the spectrum may be very different from that shown in Fig. XII-2. Figure XII-3 shows the spectrum for a solution that initially had very high photon yield, and its change with time. The shape of the spectrum changed with time due to photon intensity decrease.

The spectrum of the counts produced by a chemiluminescence reaction can be altered by variations in the resolving time. Figure XII-4 shows the spectrum for the same photon intensity at two different values of resolving time, τ_1 and τ_2, where τ_1 is less than τ_2. With the longer resolving time the probability of multiple photon events in a single pulse is greater. This will produce a greater number of pulses at the higher settings of the discriminator and will decrease the actual number of counts.

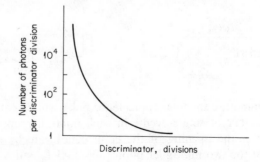

Fig. XII-2. Distribution of photon intensity from a chemiluminescence reaction.

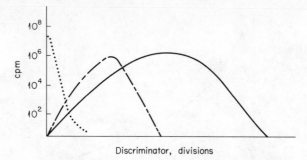

Fig. XII-3. Change in measured counts distribution of a chemiluminescence reaction with time: $\cdots t = 1$ day; $—\cdot— t = 10$ min; $—— t = 0$.

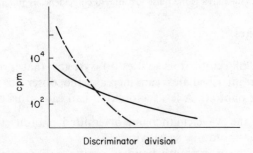

Fig. XII-4. Distribution of counts from a chemiluminescence reaction as a function of resolving time. In this experiment τ_1 ($—\cdot—\cdot—$) is less than τ_2 ($——$).

The single photoelectron events, readily measured by a single multiplier phototube, will be discriminated against in a coincidence counter. The spectrum of counts versus discriminator division setting for a single multiplier phototube, as shown in Fig. XII-2, shows that most of the counts correspond to a single photoelectron being dislodged from the photocathode; i.e., they occur at low discriminator division settings. If the same solution were observed by two multiplier phototubes with a coincidence requirement, most of the pulse at the low discriminator settings would be rejected since the probability of a coincident event occurring in the second multiplier phototube will be small. However, for a pulse occurring at the higher discriminator settings, which correspond to many photons (or photoelectrons) during the coincidence resolving time, the probability of a coincidence pulse occurring in the other multiplier phototube is high. Thus in coincidence counting mode the pulses that occur at the lower discriminator settings will decrease in number. Figure XII-5 shows a spectrum of pulse in singles and coincidence counting mode.

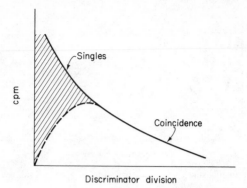

Fig. XII-5. Effect of coincidence operation on the distribution of counts from a chemiluminescence reaction. Shaded area is the pulse loss due to coincidence requirement.

Influence of Solutes

The addition of scintillator solutes to systems that produce excited
products and/or intermediates can increase the observed and even the
absolute yield of photons. A few factors that can cause this effect are:

(a) a shift of the wavelength band of emitted photons to a region of
greater multiplier phototube sensitivity,

(b) a shift of the wavelength band of emitted photons to a region of
lower absorptivity of the solvent and/or components of the solution,

(c) an increased yield of photons by acting as a scavenger of excited
molecules which themselves have very low fluorescence yields.

Figure XII-6 shows the results of the addition of varying amounts of naphthalene to a solution of Hyamine 10-X in dioxane (2). The chemilumines-

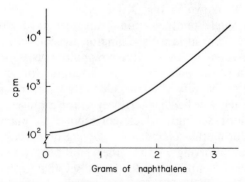

Fig. XII-6. Effect of added naphthalene on the measurement of chemiluminescence in a
dioxane solution containing Hyamine 10-X.

cence most likely is caused by a reaction between the Hyamine 10-X or impurities in the Hyamine 10-X and the dioxane producing excited states of dioxane. However, dioxane has a very low fluorescence yield, less than 0.03. The addition of increasing amounts of naphthalene increased the observed (as cpm) photon yield. The naphthalene acted as a scavenger of the excited dioxane molecules, shifted the emission spectrum to a range of greater multiplier phototube sensitivity, and emitted more photons because of its greater fluorescence yield of 0.23.

Effect of Temperature

The effect of temperature on the intensity of chemiluminescence is reflected in the temperature coefficient of the rate of the reaction that produces the excited species. Thus a reduced temperature does not eliminate chemiluminescence, as is the common belief, but only slows down the reaction rate until the number of photons per unit time (30 nsec resolving time) is low and the chance of a coincidence pulse is small.

Figure XII-7 shows the temperature effect on the chemiluminescence yield (3), measured by the coincidence count rate, for a solution of dioxane with added Hyamine 10-X. There is more chemiluminescence at the higher temperature, and the rate of change is greatest near room temperature.

Two important conclusions about the control of chemiluminescence are obtained from the data just discussed. First, if it is desirable to eliminate chemiluminescence, it can be accomplished more rapidly by increasing the temperature of the solution rather than by cooling it down. If this technique is employed, care should be exercised when dealing with volatile samples so as not to lose part of the sample activity. Second, if the count rate of the

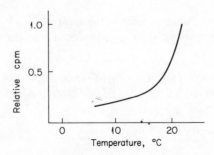

Fig. XII-7. Chemiluminescence rate as measured by observed counts as a function of the solution temperature.

sample is orders of magnitude greater than the background and chemi-luminescence, it is only necessary to be sure that the chemiluminescence intensity remains fairly constant during the counting time. To accomplish this it is necessary to maintain a constant temperature. To minimize small variations in temperature, a reduced temperature is more desirable, in the case given in Fig. XII-7, as shown by the small change in relative chemi-luminescence with change in temperature.

Detection of Chemiluminescence

There are several methods for the determination of the presence and intensity of chemiluminescence in a counting sample. Again if the sample is counted in a coincidence-type counter, the chemiluminescence rate has to be significantly high to increase the coincidence count rate very much. The chance coincidence rate equation

$$N_c = 2\tau_r N_1 N_2$$

shows that a chance coincidence rate (N_c) of 10 cpm would correspond to a singles rate of 10^5 cpm in each multiplier phototube.

Example. The noise rates of the two multiplier phototubes of a coincidence-type liquid scintillation counter are

$$PM_1 = 7000 \text{ cpm}, \qquad PM_2 = 5000 \text{ cpm}.$$

A sample containing 6000 cpm of ^{3}H was counted in coincidence. When the sample was counted in the noncoincidence mode (singles count mode) the following data were obtained:

$$PM_1 = 33,000 \text{ cpm}, \qquad PM_2 = 31,000 \text{ cpm}.$$

(a) Is chemiluminescence present in the sample?

(b) If so, what percentage error is introduced into the coincidence count rate?

(a) The singles count rate, if no chemiluminescence were present, should be approximately equal to the sum of the sample count rate and the tube noise rate:

for PM_1: $7000 + 6000 = 13,000$ cpm

for PM_2: $5000 + 6000 = 11,000$ cpm

Since both of these rates are significantly less than the measured singles rate,

for PM_1: $\quad$ 33,000 $-$ 13,000 $=$ 20,000 cpm

for PM_2: $\quad$ 31,000 $-$ 11,000 $=$ 20,000 cpm

the difference is likely due to chemiluminescence. The chemiluminescence rate of each multiplier phototube is about 20,000 cpm.

(b) The contribution at this rate of chemiluminescence to the coincidence count rate is

$$N_c = 2 \times \left(\frac{30 \times 10^{-9} \text{ sec}}{60 \text{ sec/min}} \right)(20,000)(20,000)$$

$$= 0.4 \text{ cpm.}$$

The percentage error introduced would be

$$0.4/6000 \times 100 = 0.007\% \text{ error,}$$

which is not significant.

This example points out one of the methods of determining the presence of chemiluminescence. Note the increase in singles rate of each multiplier phototube as it views the sample. If the rate is very different from the sum of the tube noise rate and sample rate, then chemiluminescence is present in the sample. Even if chemiluminescence is present, its contribution to the coincidence rate may be insignificant. Also, if the sample count rate is high, greater than 200,000 cpm, a significant amount of chemiluminescence can be tolerated, as the chance coincidence rate will be very small compared to the sample count rate. If the sample count rate is 200,000 and the chemiluminescence singles rate is 200,000, the increase in coincidence rate will be only 4 cpm.

Another technique of determining the presence of chemiluminescence utilizes the fact that the coincidence pulses can be eliminated by delaying the signal from one of the two multiplier phototubes before the coincidence-determining circuit. The delay has to be greater than the resolving time of the coincidence circuit, which is usually 20–30 nsec in most commercial instruments. Under the same conditions the chance coincidence rate is not altered since there is no true time relationship between the pulses from the two multiplier phototubes. Thus the chance coincidence rate can be directly measured by this delay technique.

A common technique used to determine the presence of chemilumines-cence is to count the sample, wait a period of time, and then recount the

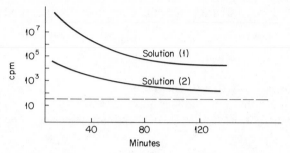

Fig. XII-8.   Decrease in measured counts due to chemiluminescence as a function of time for two different solutions: (1) Dioxane containing naphthalene (120 g/liter), PPO (4 g/liter), and POPOP (0.075 g/liter); (2) 60% toluene and 40% methyl glycol containing naphthalene (80 g/liter) and BBOT (4 g/liter). Both solutions contain additions of the same amount of basic (KOH) solution of benzoyl peroxide. The dashed line indicates the normal background level.

sample. If the count rate decreases more than the fluctuation due to statistical variations, it is used as evidence of chemiluminescence. This can lead to a false conclusion because other factors can also lead to decreases in measured count rate with time, i.e., phase separation, generation of a quencher by a chemical reaction, dissolution of oxygen into a solution, etc. However, chemiluminescence does decrease with time because the reactants are being utilized in the chemical reaction and the rate of reaction (rate of photon production) is a function of the concentration of the reactants. Figure XII-8 shows a typical plot of the chemiluminescence coincidence rate as a function of time elapsed after the sample was prepared for counting (2). The rate of decay and the initial amount of chemiluminescence depend on the chemical reactants that produce the chemiluminescence.

This technique, besides being time-consuming, requires the counting solution to be very stable over the time of investigation. If the decay requires several days, the counting solution has to remain unchanged for the total time. If the sample is volatile, extra precautions are required to ensure that none of the sample is lost.

The distribution of pulses due to chance coincidence of chemiluminescence will vary with time (2). Figure XII-9 shows the coincidence count rate due to chemiluminescence as a function of time for two counting channels, a channel from zero pulse height to the ^3H endpoint and a channel from the ^3H endpoint to the ^{14}C endpoint ($^{14}C/^3H$). This fact can be utilized to detect chemiluminescence with a two-channel counting procedure (4). Figure XII-10 shows the distributions of pulses due to ^3H beta events and those due to chemiluminescence. The ^3H efficiency is predetermined in the two counting channels, A and B. The sample is counted and the dpm is determined for each channel, dpm(A) and dpm(B). If no chemiluminescence

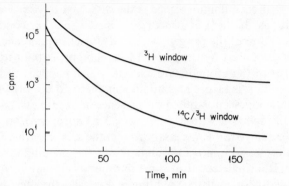

Fig. XII-9. Time decay of chemiluminescence in two counting channels measuring different pulse amplitudes.

is present, dpm(A) = dpm(B). However, when dpm(A) > dpm(B), it is very likely that chemiluminescence of the sample is producing coincidence counts.

Causes of Chemiluminescence in Liquid Scintillators

In liquid scintillation counting, the presence of chemiluminescence is undesirable. Many techniques are offered in the literature as methods for eliminating or reducing chemiluminescence. However, its presence or even its elimination is frequently mysterious, and most often unpredictable and not understood. Usually the sudden appearance of chemiluminescence can be related to a change in procedure or in a chemical used in the procedure.

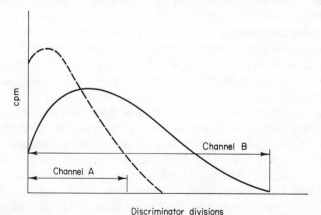

Fig. XII-10. Selection of two counting channels, A and B, to check on the possible presence of chemiluminescence. Eff(A) = Eff(B) if no chemiluminescence is present: ——— ^{3}H pulses; – – – – chemiluminescence pulses.

One source of chemiluminescence is the dissolved oxygen present in all aerated solutions. If there is also present in the solution conditions that favor the excitation of the oxygen to excited singlet oxygen, chemiluminescence will be produced by the emission of photons as the excited singlet oxygen returns to the ground-state oxygen (which happens to be a triplet state). The reaction is favored by an alkaline pH and the presence of peroxides. Chemiluminescence due to the presence of oxygen can be eliminated by flushing the solution with an inert gas, e.g., nitrogen or argon. If this technique is employed, it is necessary to seal the sample, otherwise oxygen will slowly diffuse back into the solution and the chemiluminescence will recommence. If a single count of short duration is all that is required, this technique could be used if the counting is done shortly after the inert gas flushing step. Addition of an acid to give a neutral solution and/or chemicals to react with excess peroxides is also useful in eliminating the chemiluminescence.

The choice of solvent is also important for the presence of chemiluminescence. The two major solvents used in liquid scintillation counting are toluene and dioxane, although whenever possible the use of dioxane should be avoided. Dioxane, in the past, has been used mainly for counting aqueous solutions; however, with the newly developed emulsifier systems, dioxane is finding less application. Dioxane is a cyclic diether which undergoes autooxidation in the presence of air to produce peroxides which in turn cause chemiluminescence when the dioxane is used as a solvent in a liquid scintillator. Purifying the dioxane and storage under inert atmosphere helps to some extent, but it seems only to slow down the production of the source of chemiluminescence. Besides, when the dioxane is used to prepare the scintillator solutions it usually becomes saturated with air which, in a period of time, will result in the production of chemical species (peroxides) that will cause chemiluminescence.

The purities of the solvents, solutes, solubilizers, and other chemicals added to the scintillator solution are also a factor in determining whether or not chemiluminescence will be produced. Often the source of the chemiluminescence is a reaction involving impurities or catalyzed by an impurity.

Digestion of tissues, blood, urine, and many other biological-type samples with the many basic quarternary amines commercially available often leads to a colored solution. Direct counting of the colored solution would lead to reduced counting efficiencies due to color quenching. Color bleaching agents have been used which are usually a peroxide or peroxide-producing chemical. Some commonly used chemicals are H_2O_2, benzoyl peroxide, chlorine water (HOCl), and sodium borohydride. (Sometimes activated charcoal is used to absorb the color-producing compound from the solution.)

Since excess peroxides are usually added to ensure the complete removal of all color, the direct counting of the solubilized-discolored sample is often accompanied by the presence of large amounts of chemiluminescence. The addition of a nonquenching reducing agent will usually result in reaction with the excess peroxides, and thus eliminate them as a source of chemiluminescence. In some cases it has been shown that the enzyme catalase reacted with peroxides (especially H_2O_2) to remove them as possible sources of chemiluminescence. One drawback to the use of catalase is that it releases CO_2 from an unknown origin. It is possible that $^{14}CO_2$ could be released in samples with ^{14}C, leading to a loss of sample activity.

The pH of the final counting solution is also a factor in the presence and/or intensity of chemiluminescence. Most solubilization procedures call for basic conditions. These same basic conditions are important, and often necessary, in the production of chemiluminescence. The chemiluminescence reactions involving peroxides are catalyzed by the basic pH.

In some cases simple adjustment of the pH to neutral or slightly acidic (pH ≤ 7.0) is sufficient to eliminate chemiluminescence. Acidification and/or neutralization methods depend on the nature of the sample. For aqueous samples simple additions of an inorganic acid such as HCl have been used with success. For organic soluble samples organic acids, such as ascorbic acid, have been used.

Recently it has been reported that in certain cases chemiluminescence has been observed in acidic solutions. Thus each system should be checked before it is concluded that chemiluminescence is present.

Of course one method of eliminating chemiluminescence due to solubilizing and decoloring techniques is the complete oxidation of the sample. The oxidation procedures are described in detail in Chapter VII.

Reducing Chemiluminescence

Listed below are some of the methods that have been suggested in the literature for reducing or eliminating chemiluminescence:

(a) use aromatic solvents as opposed to dioxane, whenever possible;

(b) oxidation of samples to CO_2 and H_2O by combustion;

(c) removal of oxygen from scintillator solutions by inert gas flushing or addition of oxygen getters;

(d) removal of excess peroxides (used for bleaching) by addition of reducing agents;

(e) use specially purified reagents for preparation of the scintillator solution and treatment of the sample;

(f) try to count in solutions with the pH ≤ 7.0;

(g) store samples to allow for decay of chemiluminescence (usually more rapid decay at elevated temperatures).

Bioluminescence

Recently two excellent review papers have been published on the use of liquid scintillation counters as quantum counters for the measurement of bioluminescence reactions (5, 6). The bioluminescence assay of adenosine triphosphate (ATP) by use of firefly extracts is already widely used. But the technique has also been used to assay other compounds: reduced FMN (flavinemononucleotide), NAD (nicotinamide-adenine-dinucleotide), and NADP (nicotinamideadenine-dinucleotide-phosphate). The sensitivity of this method is less than picomolar concentration (10^{-12} moles).

Bioluminescence, like chemiluminescence, is a reaction that produces an excited product which has a high probability of releasing its excess energy in the form of a photon. The process of photon production is a single photon per chemical reaction. For this reason the probability of measurement of bioluminescence is higher if the counter is operated in a noncoincidence mode; i.e., the pulses from each multiplier phototube are summed

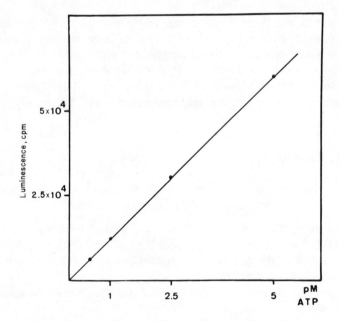

Fig. XII-11. Calibration curve showing the linear relationship between measured cpm and the log of the concentration of ATP in pico molar (10^{-12} m/liter).

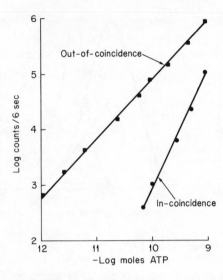

Fig. XII-12. ATP calibration curves. Out-of-coincidence discriminator settings: 60–65; amplification 100%. In-coincidence settings: 50–1000; amplification 50% (tritium channel). Using a Packard Instrument Model 3375 Tri-Carb scintillation spectrometer with high quantum efficiency EMI-9635 QB multiplier phototubes (7).

and counted. The noncoincidence operation mode will increase the background. However, the new commercially available multiplier phototubes have noise levels that are sufficiently low to allow picomolar amounts of ATP to be measured even at room temperature operation.

The bioluminescence rate will be continually changing with time, depending on the rate of the reaction. Thus, to be utilized as an analytical tool the mixing time and measuring time have to be very critical. At mixing time the bioluminescence intensity will rapidly increase and then decrease at a rate dependent on the reaction rate. For accurate assay, the mixing has to be rapid and complete as the bioluminescence rate will be determined by the concentration of the reactants.

Like any chemical reaction, the yield of products is affected by many variables. The purity of the reactants and reaction media can cause interfering reactions. The temperature, the pH, and the presence of catalysts can alter the rate of the reaction. To be sure of the method it is advisable to run a standard curve at several concentrations to determine the photon intensity (count rate) for several concentrations. Figure XII-11 shows such a plot of the concentration of ATP versus the count rate (5). The comparison of the count rate versus concentration of ATP is shown in Fig. XII-12 with coincidence and noncoincidence counting modes (7). The coincidence counting mode is useful over a very limited concentration range.

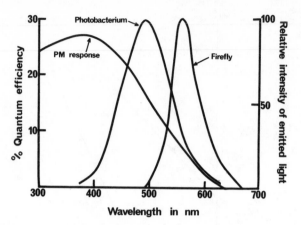

Fig. XII-13. Emission spectra of *Photobacterium* and firefly systems and spectral response of the EMI-9635 QB photomultiplier tube (7).

The useful range of ATP concentrations that give a response proportional to the concentration is 0.01–10 picomoles (6). The sensitivity is so good that it will often be necessary to dilute the samples in order to reduce the photon intensity to the level that the measured count rate is proportional to the ATP concentration. In another report (7) the useful range is claimed to extend over a much wider range, 10^{-7}–10^{-14} moles.

The specificity is very important. The firefly luciferase is specific for ATP. However, the presence of other related substances which can be converted into ATP is a source of error. A time–intensity study will often serve as a clue to the production of ATP by side reactions.

The sensitivity of this method will depend on the efficiency of the multiplier phototubes for the photon energies. The emission from biolumines-cence reactions using firefly extract is well into the red part of the visible spectrum. The sensitivity of most liquid scintillation counter multiplier phototubes to this wavelength band is quite low. The use of Photobacterium produces photons of lower wavelength and hence higher detection prob-ability. Figure XII-13 shows the typical emission spectra and the response curve for an EMI-9635 QB multiplier phototube (7).

References

1. D. M. Hercules, *in* "The Current Status of Liquid Scintillation Counting" (E. D. Bransome, Jr., ed.), p. 315. Grune & Stratton, New York, 1970.
2. D. A. Kalbhen, *in* "The Current Status of Liquid Scintillation Counting" (E. D. Bransome, Jr., ed.), p. 337. Grune & Stratton, New York, 1970.

3. D. A. Kalbhen, *in* "Liquid Scintillation Counting" (A. Dyer, ed.), Vol. 1, p. 1. Heyden, London, 1971.
4. E. D. Bransome, Jr. and M. F. Grover, *in* "The Current Status of Liquid Scintillation Counting" (E. D. Bransome, Jr., ed.), p. 342. Grune & Stratton, New York, 1970.
5. E. Schram, *in* "The Current Status of Liquid Scintillation Counting" (E. D. Bransome, Jr., ed.), p. 129. Grune & Stratton, New York, 1970.
6. E. Schram, R. Cortenbosch, E. Gerlo, and H. Roosens, *in* "Organic Scintillators and Liquid Scintillation Counting" (D. L. Horrocks and C. T. Peng, eds.), p. 125. Academic Press, New York, 1971.
7. P. E. Stanley, *in* "Organic Scintillators and Liquid Scintillation Counting" (D. L. Horrocks and C. T. Peng, eds.), p. 607. Academic Press, New York, 1971.

CHAPTER XIII

RADIOIMMUNOASSAY (RIA)

The use of radioactive tracers in immunoassay has been given the name radioimmunoassay (RIA) (1). The technique is based upon the competition between a known amount of a labeled antigen (Ag*) and an unknown amount of assay antigen (Ag) for the binding sites in a known amount of an antibody (Ab). The scheme is as follows:

$$\text{Ab} + \text{Ag} \rightleftharpoons \text{AbAg}$$

$$+$$

$$\text{Ag*}$$

$$\updownarrow$$

$$\text{AbAg*}$$

After equilibrium has been obtained the bound labeled antigen (AbAg*) and free labeled antigen (Ag*) are separated and either or both fractions are radioassayed.

258

In most RIA procedures the antigen is labeled with a radioactive gamma-emitting isotope of iodine, commonly ^{125}I and ^{131}I. These isotopes are assayed with a gamma counter. However, there are many antigens which cannot be labeled with iodine, although nearly all antigens can be labeled with ^3H and/or ^{14}C. When these radioactive tracers are used it is required that liquid scintillation counting techniques be used to measure the amount of radioactivity in the bound and/or free state. Because of the much longer half-lives of ^3H and ^{14}C (compared to ^{125}I and ^{131}I) it is necessary to use much more activity per sample to obtain the same sensitivity.

A standard amount of antibody (Ab) and radioactively tagged antigen (Ag*) is mixed with an unknown amount of the antigen (Ag). The tagged Ag* should be the same as the Ag to be assayed, or at least have the same specific reaction with the Ab. Since an equilibrium has to be established between the Ab and the Ag and Ag*, an incubation period is usually required. This time period will vary depending on the specific species involved. In some cases where the Ab–Ag are not temperature sensitive, the equilibrium can be accelerated by incubation at an elevated temperature.

After the equilibrium has been attained, it is necessary to separate the bound Ag (Ab–Ag and Ab–Ag*) from the free Ag and Ag*. There are several techniques that will accomplish this, some of which are listed below:

Liquid phase	Solid phase
Chromatoelectrophoresis	Dextran-coated charcoal
Double antibody	Antibody-coated polystyrene
Gel filtration	Antibody-coated resins
$(NH_4)_2SO_4$ precipitation	

The RIA method has to be calibrated for the particular Ab–Ag reaction which is to be monitored. This is accomplished by measuring the bound and/or free Ag* as a fraction of the amount of unlabeled Ag. Figure XIII-1 shows a typical standardization. The reaction equilibrium, separation, and measurement of ^3H in the two forms will enable the direct reading of any unknown amount of dioxin.

Many experiments are performed with the assay of only one of the Ag forms, either free or bound Ag*. The other form is assumed to be the difference from the amount added. Because some incubation periods are several days, it may be necessary to correct for the change in the amount of tracer due to the half-life (especially for ^{131}I). However, if both phases are measured at the same time, any decay effects will cancel out. There is no decay correction when the Ag is tagged with ^3H or ^{14}C.

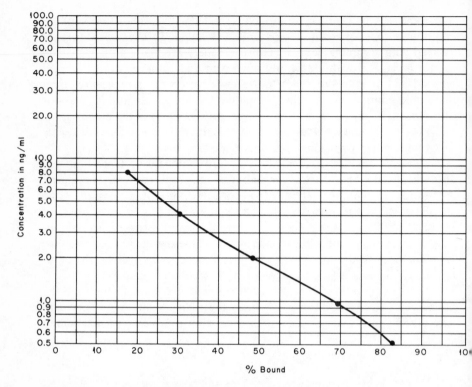

Fig. XIII-1.   Sample digoxin standard curve. A standard curve must be established for each run of unknowns. (See also Table XIII-1.)

The RIA method depends on a chemical reaction; therefore, it is necessary that precautions be taken to ensure that nothing will disturb or compete with the reaction. All material should be of the highest purity. Such factors as pH, temperature, and concentrations should be checked to determine if these have effects on the reaction. Also to be considered is the effect of the radioactive decay of the label. In this case the gamma emitter ^{131}I will induce the most radiation damage because of its high-energy emissions, both γ rays and beta particles. ^{125}I will have much less radiation damage, since it emits only low-energy X rays and γ rays. Both ^{3}H and ^{14}C decay with the emission of only beta particles of 6 and 50 keV average energy, respectively. Radiation damage can lead to by-products which could compete for binding sites on the Ab—nonspecific binding.

Of course, if liquid scintillation counting is used to assay the radioactivity, all of the problems of sample solubility, quenching, and counting efficiency will be present. However, once the separation has been made such techniques as sample oxidation can be employed to reduce or eliminate these problems.

Table XIII-1

Some sample data and sample calculations for the digoxin standard curve given in Fig. XIII-1

Sample data			
Sample data	cpm	Average cpm	Corrected cpm
Background	154.9	157.8	—
	160.7		
Standard 0 (100%)	2221.1	2220.7	2062.9
	2219.3		
Duostandard A	1873.7	1872.2	1714.4
	1870.7		
Duostandard B	1571.7	1585.4	1427.6
	1599.1		
Duostandard C	1161.5	1162.4	1004.6
	1163.3		
Duostandard D	782.5	782.7	624.9
	782.9		
Duostandard E	509.7	516.0	358.2
	522.3		

Sample calculations

$$\% \text{ Bound }^3\text{H digoxin at } 0.5 \text{ ng/ml} = \frac{1714.4}{2062.9} \times 100 = 83.1\%$$

$$\% \text{ Bound }^3\text{H digoxin at } 1.0 \text{ ng/ml} = \frac{1427.6}{2062.9} \times 100 = 69.2\%$$

$$\% \text{ Bound }^3\text{H digoxin at } 2.0 \text{ ng/ml} = \frac{1004.6}{2062.9} \times 100 = 48.7\%$$

$$\% \text{ Bound }^3\text{H digoxin at } 4.0 \text{ ng/ml} = \frac{624.9}{2062.9} \times 100 = 30.3\%$$

$$\% \text{ Bound }^3\text{H digoxin at } 8.0 \text{ ng/ml} = \frac{358.2}{2062.9} \times 100 = 17.4\%$$

It is also possible to use liquid scintillation techniques to assay the ^{131}I and ^{125}I. Both of these nuclides can be counted in a liquid scintillator solution with $\sim 89\%$ efficiency, and with $\sim 75\%$ efficiency by counting the betas and electrons with liquid scintillator solutions.

Several different commercial concerns have made available on the market many RIA kits which supply all the required materials to perform the standardization curve and several assays. The sensitivity is claimed to be in the range of $10^{-9}–10^{-12}$ g of the Ag, but the purity is very critical at these lower limits. A major problem in RIA is the preparation of the labeled Ag with the high purity required.

Figure XIII-1 shows a radioimmunoassay test for digoxin levels in blood (serum) samples. These data were taken as an example only. The test is from a kit manufactured by Kallestad Laboratories, Inc., Chaska, Minnesota. In this procedure the bound and free digoxins are separated by absorption of the free digoxin on dextran-coated charcoal. The supernatant (~ 1.2 ml) is decanted into a liquid scintillation vial and emulsified in a liquid scintillation solution containing an emulsifier (i.e., Triton X-100). The samples are counted in a liquid scintillation counter. It is necessary to measure the quench level of each sample, as different patient serums have different impurities. The quench correction is also important because the standard curve may be run with samples having a different quench level than the patient samples. Sample data and calculations for the standard curve in Fig. XIII-1 are given in Table XIII-1.

Reference

1. R. S. Yalow and S. A. Berson, *Nature (London)* **184,** 1648 (1959).

CHAPTER XIV

CERENKOV COUNTING

Many investigators are now using liquid scintillation counters to measure the Cerenkov radiation (1, 2) produced by high-energy electrons. No scintillators are required for the production of Cerenkov radiation. The only requirement is that the energy of the electron be greater than a threshold energy (dependent on the solvent). However, most of the Cerenkov radiation is in the ultraviolet wavelength region, where the efficiency of the MPTs of the liquid scintillation counter is quite low. Some new wavelength shifters have been developed to increase the photon yield in the wavelength region of high MPT sensitivity. The first measurement of radioactivity by detection of Cerenkov radiation in a liquid scintillation counter was reported by DeVolpi and Horrocks in 1961 (3). They counted aqueous solutions of $^{56}MnSO_4$ with high efficiencies, 90% in a single MPT system and 35% in a coincidence MPT system.

Cerenkov Effect

As charged particles pass through a medium, there can be an exchange of energy from the charged particle to the molecules of the medium. The exchange energy produces local electronic polarizations along the path of the charged particle. As these polarized molecules return to their normal state the excess energy is released as photons. The intensity distribution of the photons is shown in Fig. XIV-1. The relative number of Cerenkov photons will decrease as λ^{-3} as a general rule.

A velocity threshold exists for the production of Cerenkov radiation. The threshold is a function of the refractive index of the medium n. If the velocity of the particle v is greater than the velocity of light c divided by n ($v > c/n$), Cerenkov radiation will occur and will be directional in nature. The half-angle ϕ of the cone of emitted radiation is given by the relationship:

$$\cos \phi = c/vn$$

and the threshold energy E_{min} for production of Cerenkov radiation is given by the relationship:

$$\beta_{min} = 1/n$$

where

$$\beta = v/c \quad \text{and} \quad E = 0.511((1 - \beta^2)^{-1/2} - 1).$$

Figure XIV-2 shows the minimum energy for Cerenkov radiation production (i.e., energy threshold) as a function of the refractive index of the medium. The threshold energy can be calculated from the equation:

$$E_{min} = 0.511((1 - 1/n^2)^{-1/2} - 1) \text{ MeV}$$

where 0.511 is the rest mass of an electron in million electron volts.

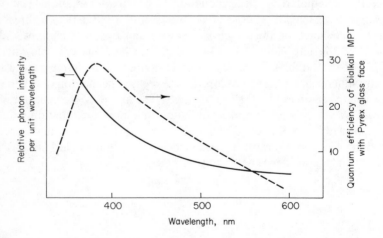

Fig. XIV-1.   Typical photon distribution from Cerenkov radiation and typical response of a bialkali MPT.

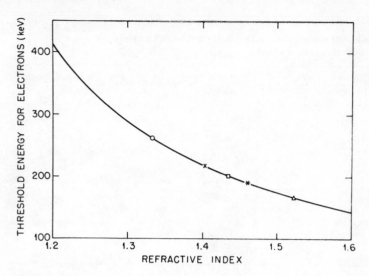

Fig. XIV-2. The threshold beta energy for Cerenkov radiation as a function of the refractive index of the medium through which the betas are passing. ○ H_2O; × glycol 56% (wt.); □ glycerol 75% (wt.); * glycerol 95% (wt.); △ sucrose 85%.

Cerenkov Counting Efficiency

Only those beta particles with energies greater than the threshold energy have a probability of being counted. Thus the maximum theoretical efficiency can be calculated from beta energy spectrum. Table XIV-1 lists a few typical beta emitters, their E_{max}, and the fraction of the betas above 263 keV, which is the E_{min} for water.

There is a large difference, especially at low energies, between the fraction of beta particles which produce Cerenkov radiation and the counting efficiency. This is in part due to the facts that

(a) most of the radiation is of wavelengths for which the MPT is insensitive, and

(b) the radiation is directional so that in a coincidence counter the probability of a coincidence event is reduced.

These two factors are partly overcome by the use of wavelength shifters (4–10). These compounds absorb the Cerenkov radiation and reradiate the energy as photons of longer wavelength in an isotopic pattern. Thus by proper choice of wave shifter, the photon wavelengths will match the maximum sensitivity of the MPT. The isotopic emission increases the probability of a true coincidence for a given beta event. Table XIV-2 shows

Table XIV-1

Fraction of beta particles from several radionuclides which exceed the Cerenkov threshold of water[a]

Isotope	E_{max} (MeV)	Fraction of beta spectrum above 263 keV
[3]H	0.018	0
[14]C	0.158	0
[45]Ca	0.254	0
[131]I	0.815 (0.7%)	
	0.608 (87.2%)	0.31
	0.335 (9.3%)	
	0.250 (2.8%)	
[36]Cl	0.714	0.46
[90]Sr–[90]Y	0.545	0.61
	2.26	
[32]P	1.71	0.86

[a] 263 keV.

some data by Parker (11), which give the counting efficiency with and without wave shifter (double sodium potassium salt of 2-naphthylamine-6, 8-disulfonic acid at a concentration of 100 g/liter).

Volume Effect

Several investigators have observed an effect of the volume of the aqueous solution on the counting efficiency by Cerenkov counting (6, 11, 12). Figure XIV-3 shows the volume dependency obtained (11). This effect will vary, depending on the type of vial used and the geometry and optics of the sample counting chamber and MPTs of the counter. Keep in mind that the detection limit may be improved by using the maximum volume of the vial, since the detection limit will be the product of volume times efficiency.

In a study of effects of volume (13), it was noted that the relative counting efficiency of lower energy beta emitters varied more with volume. Figure XIV-4 shows results with three different energy beta emitters: [185]W (E_{max} = 0.427 MeV), [36]Cl (E_{max} = 0.714 MeV), and [32]P (E_{max} = 1.710 MeV). The necessity of maintaining constant volume is obvious from these results.

Background

The background is independent of the volume of the aqueous solution, assuming there is no natural radioactivity in the water—[226]Ra, [232]Th, etc.

Table XIV-2

Cerenkov counting efficiencies determined with different types of multiplier phototubes and with and without a wavelength shifter[a]

| | | Counting efficiencies (% of dpm) | | |
| | | S-11 | | Bialkali |
Isotope	E_{max} (MeV)	Aqueous	Aqueous with shifter[b]	Aqueous
^{36}Cl	0.71	2.3	4.7	5.3
^{40}K	1.32	14	31	34
^{24}Na	1.39	18	40	
^{32}P	1.71	25	50	43
^{42}K	2.0, 3.6	60	85	76

[a] See R. P. Parker and R. H. Elrick, *in* "The Current Status of Liquid Scintillation Counting" (E. D. Bransome, Jr., ed.), p. 110. Grune & Stratton, New York, 1970.

[b] Double sodium potassium salt of 2-naphthylamine-6,8-disulfonate at 100 g/liter.

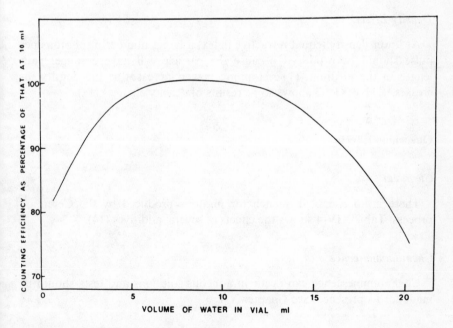

Fig. XIV-3. The relative Cerenkov counting rate as a function of the amount of water in the counting medium (11).

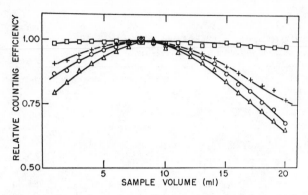

Fig. XIV-4. Effect of volume of counting medium on relative Cerenkov counting for different energy beta emitters. (□) ^{14}C, 0.157 MeV (in liquid scintillation solution); (△) ^{185}W, 0.427; (○) ^{36}Cl, 0.714; (+) ^{32}P, 1.710.

Most glass vials will give backgrounds of about 25–30 cpm, while plastic vials will give 15–20 cpm backgrounds (see Chapter XIII on vials and Chapter IX on background).

Density Effect

At lower density (equal refractive index) a larger number of photons per pulse should be produced because the particle will have a longer path length in the medium. (The stopping power increases as the density increases.) Table XIV-3 shows some results of density changes (11).

Quenching Effects

Chemical

There is no chemical quench for photons produced by the Cerenkov process. Table XIV-4 shows the effects of several additives (14).

Chemiluminescence

Chemiluminescence can occur in aqueous solutions and tests should be made for its presence. See Chapter XII.

Color Quench

Additives that absorb light in the ultraviolet and near-ultraviolet wavelength regions will lead to reductions in the number of photons that escape

Table XIV-3

Effect of density of the Cerenkov medium on counting efficiency[a]

Isotope	E_{max} (MeV)	Density change (%)	Observed increase in counting efficiency (%)
^{204}Tl	0.77	7	10
^{32}P	1.710	7	1

[a] See R. P. Parker and R. H. Elrick, *in* "The Current Status of Liquid Scintillation Counting" (E. D. Bransome, Jr., ed.), p. 110. Grune & Stratton, New York, 1970.

the sample vial. This type of quenching process will be related to the extinction coefficient of the quench material in the Cerenkov emission wave band and the concentration of that material. The effect can be predicted by the Lambert–Beer law at low to moderate concentrations. In Table XIV-5 are listed additives that have absorption properties in the Cerenkov

Table XIV-4

Effect of uncolored additives on the Cerenkov counting rate[a]

Additive (concentration in %)	Count rate relative to count rate prior to addition of additive to volume of	
	5 ml	10 ml
$NaNO_3$ (20)	1.00	0.91
$Ca(NO_3)_2$ (10)	0.91	0.91
HNO_3 (6.5)	1.00	0.88

	Deviation (%)
NaCl (5)	<5
Na_2SO_4 (5)	<5
Na_2CO_3 (5)	<5
Na_2HPO_4 (5)	<5
NH_4Cl (5)	<5
$SrCl_2$ (5)	<5
$BaCl_2$ (5)	<5
$ZnCl_2$ (5)	<5
$Al_2(SO_4)_3$ (5)	<5
HCl (1 N)	<5
H_2SO_4 (6)	<5
CH_3COOH (3)	<5
NH_3 (5)	<5
NaOH (30)	<5

[a] See V. K. Haberer, Packard Tech. Bull. No. 16. Packard Instrum. Co., Inc., Downers Grove, Illinois, 1966.

Table XIV-5

Effect of colored additives on the Cerenkov counting rate[a]

Additive (concentration in %)	Count rate relative to count rate prior to addition of additive to volume of	
	5 ml	10 ml
MnSO$_4$ (10)	91	88
CuSO$_4$ (10)	51	58
[Cu(NH$_3$)$_4$]SO$_4$ (?)	40	35
CuCl$_2$ (10)	37	37
K$_3$[Fe(CN)$_6$] (?)	14	12
FeCl$_3$ (10)	5	3

[a] See V. K. Haberer, Packard Tech. Bull. No. 16. Packard Instrum. Co., Inc., Downers Grove, Illinois, 1966.

emissions region (14). Those chemicals that appear yellow show a particularly strong quenching effect (due to strong blue absorption).

Haberer (14) studied the effect of picric acid concentration on the relative Cerenkov counting rate of ^{40}K aqueous solution. The relationship of picric acid concentration and the extinction is $E = \log R_0/R$, where E is the extinction fraction, R_0 the rate with no added picric acid, and R the rate at the concentration noted. The results are shown in Fig. XIV-5. The relationship was linear with respect to the logarithm of picric acid concentration. Thus the extinction of Cerenkov intensity can be used to determine the concentration of picric acid.

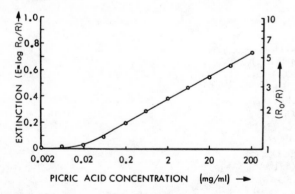

Fig. XIV-5. Relationship between picric acid (color quench) concentration and the extinction (E) of the relative Cerenkov counting rate where R_0 is the rate with no added picric acid and R is the rate at the given picric acid concentration (14).

Color Quench Correction

Several methods are used to correct the measured counting rate for color quench of the Cerenkov radiation. Some of these methods are listed in Table XIV-6.

Wavelength Shifters

Several compounds have been utilized as wavelength shifters of Cerenkov radiation as a means of increasing the detection efficiency in coincidence-type liquid scintillation counters. These wavelength shifters shift the photon energies from a region of low MPT sensitivity to a region of high MPT sensitivity. They also eliminate the directional properties of the Cerenkov radiation and produce an isotopic photon emission, which increases the probability of a coincidence event, which is required before a count will be recorded.

Ross (9) tested 12 compounds as wavelength shifters in aqueous solutions. Some of the results are given in Table XIV-7. Most of these compounds have been used as fluorescent pH indicators. Several of the compounds

Table XIV-6

Methods of correcting the color quenching of Cerenkov counting

Method	Comments
Decoloration	Use chemicals to destroy color. Change oxidation state of metal ion to noncolored species
Internal standard	Add small, known amount of nuclide and measure efficiency. Cannot recount sample. However, need not be in the same form as sample as long as no new color quenching is added
Spectroscopic	Construct plot of % transmission at given λ vs. counting efficiency. % transmission of true sample will give counting efficiency from graph
Channels ratio	The pulse height distribution of Cerenkov light pulses is similar to a beta spectrum, and the choice of two channels will allow for construction of a channels ratio vs. counting efficiency curve
External standardization	Using a high-energy γ-ray source, part of the Compton scattered electrons will exceed the Cerenkov threshold. The efficiency vs. external standard counts, or channels ratio curve, can be obtained from samples of known dpm

showed definite pH effects. The relative counting efficiency was· usually lower in solutions with pH above 7.

The wavelength shifters showed a greater effect on counting efficiency for lower energy beta emitters. An increase of a factor of ~ 4 was measured for ^{36}Cl, whereas only an increase of a factor of ~ 1.5 was measured for ^{32}P.

The concentration of the wave shifter is important. Normally concentrations of about 100 g/liter will give the maximum counting efficiency. Figure XIV-6 shows a plot of wave shifter concentration versus relative counting efficiency (9).

Some of the wave shifters showed chemical decomposition over several hours. This led to the production of color quenchers in some cases, which

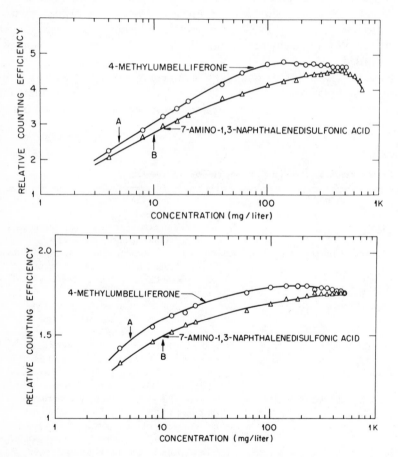

Fig. XIV-6. Effect of wavelength shifter concentration on relative Cerenkov counting rate for ^{36}Cl (a) and ^{32}P (b) (9).

Table XIV-7

Effect of wavelength shifter on the relative counting efficiency of $^{90}Sr-^{90}Y$ in single or coincidence counting mode[a]

Compound	Relative counting efficiency[b]	
	Coincidence	Single
α-Naphthol	1.00	1.00
β-Naphthol	1.23	1.19
1-Naphthylamine	1.37	1.36
Coumarin	0.96	0.97
Acridine	1.09	1.08
4-Methylumbelliferone	1.44	1.41
2′,7′-Dichlorofluorescein	(1.07)[c]	[c]
Fluorescein	(1.15)[c]	[c]
5-Amino-2,3-dihydro-1,4-phthalazinedione	(1.25)[c]	[c]
2-Naphthol-3,6-disulfonic acid (Na)	1.40	1.34
7-Amino-1,3-naphthalenedisulfonic acid	1.45	1.39
2-Naphthol-6,8-disulfonic acid (K)	1.41	1.36

[a] See H. H. Ross, *in* "Organic Scintillators and Liquid Scintillation Counting" (D. L. Horrocks and C. T. Peng, eds.), p. 757. Academic Press, New York, 1971.
[b] Relative to 1.00 for α-naphthol in each mode.
[c] Large amounts of phosphorescence present.

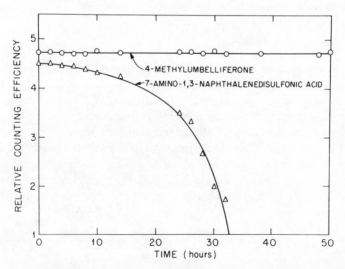

Fig. XIV-7. Relative Cerenkov counting rate of ^{36}Cl with two different wave shifters as a function of time. The decreased rate with time (7-amino-1,3-naphthalenedisulfonic acid) is due to radiation decomposition of the wave shifter (9).

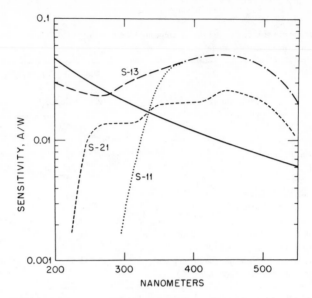

Fig. XIV-8. Response function of MPTs with different types of faces on the photocathode shown relative to the Cerenkov photon yield (solid line). S-11: Cs–Sb/glass; S-13: Cs–Sb/silica; S-21: Cs–Sb/uv glass (9).

caused a decrease in the counting efficiency with time. Figure XIV-7 shows this effect with 7-amino-1,3-naphthalenedisulfonic acid (9).

Effect of Multiplier Phototube

The newer bialkali photocathode MPTs showed a higher counting efficiency for most Cerenkov radiation because their photosensitivity is toward the lower wavelength region where the intensity of Cerenkov radiation is greater. Also, the use of quartz-faced MPTs and quartz vials will give higher counting efficiencies, because the quartz will pass more of the near-ultraviolet Cerenkov radiation. The pyrex-type glasses start absorbing photons at 3400 Å, and the absorption is essentially complete for photons of wavelengths less than 3000 Å. Figure XIV-8 shows typical response functions of several types of MPTs relative to the Cerenkov photon distribution.

References

1. P. A. Cerenkov, *Dokl. Akad. Nauk SSSR* **2**, 451 (1934).
2. L. Mallet, *C. R. Acad. Sci. Paris* **183**, 274 (1926).
3. A. DeVopli and D. L. Horrocks, Unpublished results, 1961; First reported by D. L. Horrocks, Packard Tech. Bull. 3. Packard Instrum. Col., Inc., Downers Grove, Illinois, 1961.

4. E. Heilberg and J. Marshall, *Rev. Sci. Instrum.* **27,** 618 (1956).
5. N. Porter, *Nuovo Cimento* [10]. **5,** 526 (1957).
6. V. K. Haberer, *Atomwirt. Atomtech.* **10,** 36 (1965).
7. R. H. Elrick and R. P. Parker, *Int. J. Appl. Radiat. Isotop.* **19,** 263 (1968).
8. G. Cosme, S. Jullian, and J. Lefrancois, *Nucl. Instrum. Methods* **70,** 20 (1969).
9. H. H. Ross, *in* "Organic Scintillators and Liquid Scintillation Counting" (D. L. Horrocks and C. T. Peng, eds.), p. 757. Academic Press, New York, 1971.
10. A. Läuchli, *in* "Organic Scintillators and Liquid Scintillation Counting" (D. L. Horrocks and C. T. Peng, eds.), p. 771. Academic Press, New York, 1971.
11. R. P. Parker and R. H. Elrick, *in* "The Current Status of Liquid Scintillation Counting" (E. D. Bransome, Jr., ed.), p. 110. Grune & Stratton, New York, 1970.
12. T. Clausen, *Anal. Biochem.* **22,** 70 (1968).
13. H. H. Ross, *Anal. Chem.* **41,** 1260 (1969).
14. V. K. Haberer, Packard Tech. Bull. No. 16. Packard Instrum. Co., Inc., Downers Grove, Illinois, 1966.

CHAPTER XV

PULSE SHAPE DISCRIMINATION

The intensity of photons from a scintillation event as a function of time can be divided into at least two components, fast (or prompt) and slow. (Figure XV-1 shows a typical intensity–time measurement.) The fast component has a decay time equal to the fluorescence lifetime of the fluorescing species (the fluorescence solute). The slow component has a decay time which is diffusion controlled; at room temperature in an aromatic solvent the decay time is about 200–300 nsec (1).

The total emission intensity–time equation as a function of the decay time τ_n of each component is given by the expression (2, 3)

$$I(t) = I_1 \exp(-t/\tau_1) + I_2 \exp(-t/\tau_2) \cdots I_n \exp(-t/\tau_n).$$

Normally only the two decay components are observed, $\tau_1 \sim 1\text{--}3$ nsec and $\tau_2 \sim 200\text{--}300$ nsec.

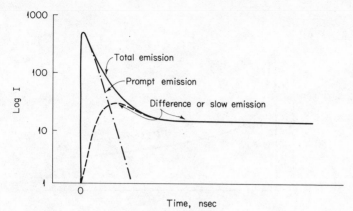

Fig XV-1. A representation of the scintillation intensity from an organic scintillator solution as a function of time showing the division of the total intensity into the prompt and slow components.

The diffusion-controlled component (i.e., slow) has been shown to be due to the annihilation of two triplet excited molecules (4–10):

$$T + T \rightleftarrows S_1 + S_0.$$

The energy transfer process results in the formation of a ground-state molecule S_0 and a singlet by excited molecule S_1. The S_1 molecule will fluoresce with the emission of photons which are the same energy distribution as those produced directly (i.e., prompt):

$$S_1 \rightarrow S_0 + h\nu.$$

Since the slow component is the result of triplet excited molecules, the absence of any quenchers, especially oxygen, is a requirement for the ob-

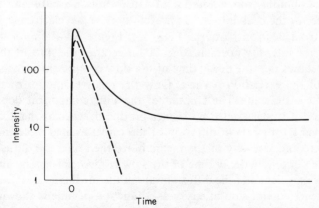

Fig. XV-2. The effect of oxygen on the scintillation intensity showing the quenching of the slow component: —— absence of O_2; ---- saturated with O_2.

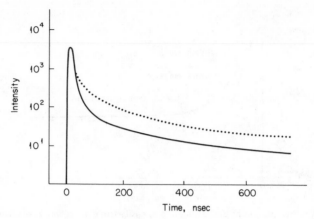

Fig. XV-3. Scintillation intensity (normalized at the peak intensity) for excitation by electrons and protons, showing difference in relative amounts of slow component: ——— electron, · · · · recoil proton.

servance of the slow component. Figure XV-2 shows the effect of oxygen on the pulse intensity as a function of time. There is very little effect on the prompt component, whereas the slow component has been completely removed. Most quenches will preferentially quench the triplet states due to their longer life before annihilation and corresponding greater probability of interacting with a quench molecule (11, 12).

The production of triplet excited molecules is related in part to the specific ionization of the excitation particle (4). Two different types of particles which produce the same intensity of the fast component will have different intensities of the slow component. Figure XV-3 is a plot of the intensity of photons from a liquid scintillator solution as a function of time (normalized at the peak intensity) for excitation by fast electrons and recoil protons (from neutron scattering) (13–15). Figure XV-4 shows the same data with the intensity normalized at 200 nsec after the start of the pulse. This plot shows that the decay time of the slow component is the same regardless of the excitation particle. Only the relative amount of the slow component is different. The amount of the slow component depends on the number of triplet excited molecules produced, whereas the decay time is controlled by the rate of diffusion of the triplet excited molecule. If the temperature and viscosity of the medium are the same, the diffusion rate will be the same. The decay time of the slow component will be altered by those factors which alter the diffusion rate of the medium.

The integrated emission intensity as a function of time is shown in Fig. XV-5 for excitation by an electron and a proton which produced the same total integrated intensity. In the first 50 nsec, the electron pulse had reached

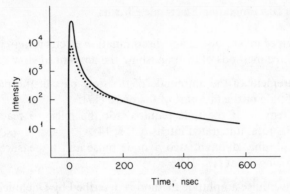

Fig. XV-4. Data from Fig. XV-3 but normalized at 200 nsec after the start of the scintillations: —— electron; · · · · recoil proton.

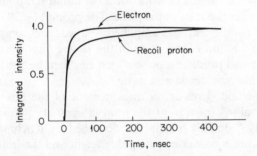

Fig. XV-5. Integrated scintillation intensity plot showing the difference due to the type of particle producing excitation. Intensities normalized at total intensities.

95% of its total integrated intensity, whereas the proton pulse had reached only 80% of its total integrated intensity. After 100 nsec the fraction of the total integrated intensities is 97% for electrons and 85% for protons (14).

The main use of pulse shape discrimination is to distinguish two types of particles with different specific ionization which are exciting the liquid scintillator at the same time. It is possible to distinguish which pulses are the result of electron excitations in the presence of recoil protons or alpha particles (16). This technique has great potential in many radiation survey and counting situations. One of these is monitoring for the individual contribution of neutrons and γ rays to the background around a reactor. Another special application is the determination of the beta and alpha contributions to the count rate of an environmental sample of very low count rate.

Pulse Shape Discrimination Electronic Circuits

The design of most pulse shape discrimination electronic circuits is based on one of three methods of distinguishing the amount of slow component:

1. measurement of the amplitude of the slow part of the pulse relative to the amplitude of the fast part of the pulse (17);
2. measurement of the time required for the pulse to reach a certain fraction of its total integrated intensity (13, 14).
3. use of doubly differentiated dynode pulse and a measure of time of the zero crossing point (18–21).

The relative pulse amplitude method is described by Daehnick and Sherr (17). The pulse is divided at a selected time after the start of the pulse. The division is usually set after the decay of almost all of the prompt component but before appreciable decay of the slow component. A compromise time for this division which seems to work for most liquid scintillator solutions is about 90 nsec after the start of the pulse. The prompt (<90 nsec) and the slow (>90 nsec) components are amplified, shaped, and compared. Those having a greater fraction of the pulse in the slow component than a predetermined value will produce a pulse which can then be used to initiate a coincidence or anticoincidence gate pulse.

The second method depends on measurement of the time required to reach a predetermined fraction of the total integrated intensity. The electronic circuit based on this principle is described by Kuchnir and Lynch (14). The time value is converted to a pulse height and the time selection is made on a simple single channel pulse height analyzer. Figure XV-6 shows the distribution of counts per unit time as a function of time. The time was

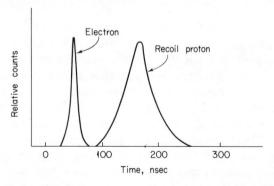

Fig. XV-6.   Relative number of counts as a function of time required for integrated intensity to reach 0.89 of the total intensity.

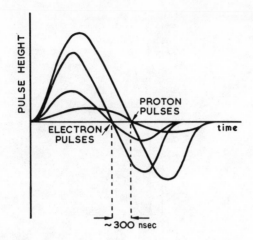

Fig. XV-7.   Idealized shape of the doubly differentiated dynode pulses.

selected at which the pulse reached 0.89 of the total integrated maximum value. Electron pulses reached this value in 50 nsec, whereas pulses from recoil protons required 180 nsec to reach the same value of their maximum. By setting a discriminator to accept pulse from recoil protons only, the resulting pulse can be used to initiate a coincidence or anticoincidence gate pulse.

The zero crossing technique depends on the shape of the doubly differentiated dynode pulses having a given crossing time dependent on the type of exciting particle. Figure XV-7 shows a representation of idealized pulse shapes for this technique. The difference in the time of the crossover can be used to trigger a coincidence or anticoincidence gate.

Performance

Each of these techniques has been shown to work under real conditions. One major limitation is the ability to differentiate between the types of particles when the energy of the particles is low. Figure XV-8 shows how the separations of the times required to reach 0.89 of the integrated pulse intensity have more overlap as the energy decreases.

The actual peak time is unchanged, i.e., 50 nsec for electrons and 180 nsec for recoil protons. However, it becomes more difficult to separate pulses due to one or the other type of particles as the energy of the exciting particle decreases. There is actually about a 10:1 energy ratio for equal

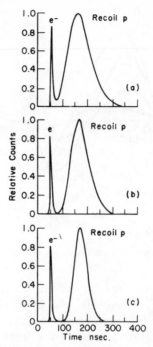

Fig. XV-8. Distribution of counts per unit time as a function of the time required for the integrated intensity of the scintillation to reach 0.89 of its maximum value for three different total integrated intensities. The resolution is poorer at lower energies (lower total integrated intensity) of the exciting particles. Total intensity equivalent: (a) ~ 35-keV electron, ~ 350-keV recoil proton; (b) ~ 50-keV electron, ~ 500-keV recoil proton; (c) ~ 75-keV electron, ~ 750-keV recoil proton.

response between recoil protons and electrons (22). Table XV-1 lists the fraction of electron pulses that are counted along with the proton pulses as a function of the fraction of the total proton spectrum which is accepted (14). The separation is shown at two different neutron energies.

Pulse shape discrimination can be done with liquid, plastic, and organic crystal scintillators provided that there are no quenches present in the scintillator (23, 24). The ratio between the slow component pulse height from electrons and protons varies with the type of scintillator. Some typical scintillators and the ratios are given in Table XV-2.

Applications

The pulse shape discrimination technique is used to detect recoil protons, alpha particles, fission fragments, or other charged particles with high

Table XV-1

The fraction of recoil protons (from neutron interactions) counted by pulse shape discriminator methods and the fraction of electrons (from scattered gamma rays) which are not rejected[a]

Neutron energy (keV)	Fraction of recoil protons counted (%)	Fraction of γ rays counted under same conditions (%)
350	90	1.37
	80	0.97
	70	0.70
500	90	0.20
	80	0.14
	70	0.11

[a] See F. T. Kuchnir and F. J. Lynch, *IEEE Trans. Nucl. Sci.* **NS-15**, 107 (1968).

Table XV-2

The ratio of the amplitudes of the slow components of electron and proton pulses for different types of scintillators

Scintillator	R^a
Crystal	
Stilbene	1.8
Anthracene	2.1
Liquid	
Toluene + PPO (5 g/liter) + M_2-POPOP (0.1 g/liter)	1.8
NE-211[b]	1.6
Plastic	
NE-150[b]	1.1

[a] R = (amplitude of slow component of proton)/(amplitude of slow component of electron).
[b] Nuclear Enterprises, Ltd.

specific ionization in the presence of electrons, or vice versa. Discrimination between two types of the high specific ionization particles is more difficult (23).

Figure XV-9a shows the pulse height distribution of the total integrated emission of a liquid scintillator solution excited by Compton scattered electrons from a source of γ rays and by alpha particles dissolved in the scintillator solution. Using the pulse shape discriminator pulse in coincidence, Fig. XV-9b, and in anticoincidence, Fig. XV-9c, the pulses due only to alpha particles or electrons, respectively, are measured. The pulse shape discrimination circuit puts out a pulse for every alpha particle which it detects by the slow pulse amplitude. This pulse was used to allow only the desired pulse to be analyzed and/or counted.

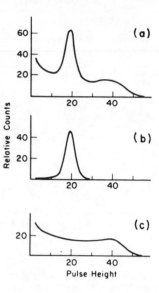

Fig. XV-9. Relative pulse height spectra from a liquid scintillator solution which was excited by electrons and alpha particles showing the use of the pulse shape discriminator (PSD) in the coincidence mode to reject electrons and in the anticoincidence mode to reject alpha particles. Alpha particles plus γ rays: (a) without PSD; (b) with PSD in coincidence; (c) with PSD in anticoincidence.

Pulse shape discrimination has been used frequently to measure the fraction of the background around a reactor which is due to γ rays or to neutrons (11, 16, 22, 25–27). The γ rays are measured by production of Compton scattered electrons in the liquid scintillator, and the neutrons are measured by production of recoil protons in the liquid scintillator. Thus the pulse shape discrimination circuit is distinguishing between electrons and protons.

Figure XV-10 shows some typical data using pulse shape discrimination to distinguish between neutrons and electrons. The top line is due to the simultaneous excitation of the liquid scintillator by sources of γ rays and neutrons. The dotted line is obtained by excitation of the liquid scintillator

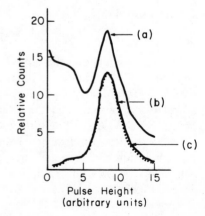

Fig. XV-10. Spectra showing the use of a pulse shape discriminator to reject pulses produced by electrons while accepting pulses produced by captured neutrons in a boron-loaded liquid scintillator solution: (a) γ rays + neutrons (without PSD); (b) solid line—γ rays + neutrons (with PSD); (c) dotted line—neutrons alone (without PSD).

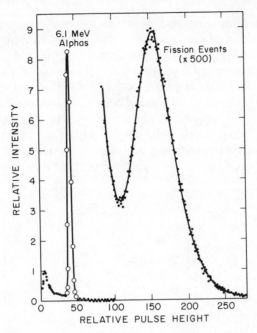

Fig. XV-11. Pulse height distribution for source of ^{252}Cf dissolved in liquid scintillation solution *without* pulse shape discrimination.

with the neutron source by capture of the neutrons by ^{10}B in the scintillator solution. The bottom solid line was obtained using the pulse shape discrimination pulse as a gate to allow only those pulses which are due to neutron interactions when the liquid scintillator is excited by both the γ ray and neutron sources. The spectrum with pulse shape discrimination is identical with that obtained for neutrons alone. This shows how well the technique works.

Pulse shape discrimination has been used to distinguish between pulses due to electrons (beta particles) and those due to fission fragments (23). A source of ^{252}Cf was dissolved in a liquid scintillator. This source produced alpha particles, fission fragments, and electrons from the subsequent decay of the radioactive fission fragments. Figure XV-11 shows the pulse height spectrum obtained in a normal liquid scintillation counter. The isotope ^{252}Cf decays by both alpha emission and spontaneous fission. One nuclear property which is desirable to know is the ratio between these two modes of decay. From the spectrum shown in Fig. XV-11 these data can be obtained by summing the number of counts under each of the two peaks in the pulse height distributions and calculating the alpha-to-fission ratio. However, because of pulses from electron excitations occurring at the same pulse heights as those from the alpha particles and fission fragments, corrections would have to be made to obtain the true counts due only to the alpha particles and fission fragments. The correction is more critical for determination of the fission fragment counts.

Use of the pulse shape discrimination technique gave the results shown in Fig. XV-12. The number of counts due to electron excitations has essentially been eliminated. The log scale of counts shows that electron-produced pulses in the region between the alpha peak and the fission peak have been reduced to ≤ 2 counts/unit pulse height. The total counts due to alpha particles and fission fragments are equal to the disintegration rate for decay by that particular decay mode since both of these excitations are counted

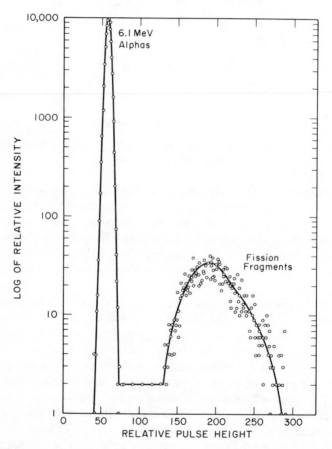

Fig. XV-12. Pulse height distribution of pulses in coincidence with the output of the PSD circuit which was set to reject pulses produced by electrons from the ^{252}Cf source dissolved in a liquid scintillation solution.

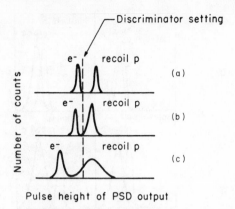

Fig. XV-13. Pulse shape discriminator output and resolution variations with different energy equivalents of the exciting particles showing loss of discrimination ability with lower-energy events. Energy equivalent: (a) ~1.5-MeV electron, ~15-MeV recoil proton; (b) ~0.6-MeV electron, ~6-MeV recoil proton; (d) ~0.06-MeV electron, ~0.6-MeV recoil proton.

with 100% efficiency for unquenched or slightly quenched solutions. The calculated alpha-to-fission ratio from the data in Fig. XV-12 is 31.0 ± 0.5, which compares to the value of 31.2 ± 0.3 obtained by counting with gas proportional counters (23).

Limits to Pulse Shape Discrimination

Efficient pulse shape discrimination is only possible when the statistical fluctuation in the measured pulse height is small compared to the pulse height value. In the case of low-energy events the fluctuation becomes a significant amount of the total pulse height, and separation of pulses due to different types of particles becomes difficult. Figure XV-13 shows the pulse shape discriminator circuit output at different energies (the response of the liquid scintillator was such that the energy ratio between electrons and recoil protons which gave equal total response was 1:10, respectively) (16).

As the energy decreases, the spectra become broader and the ability to separate pulses due to one type of particle becomes more difficult. If the discriminator settings are selected using high-energy particles, some low-energy recoil proton events will be counted as electrons. One important research effort at present is to extend the low-energy limit of pulse shape discriminations.

The choice of different compositions of liquid scintillator solutions can affect the ability to discriminate between different types of particles (29).

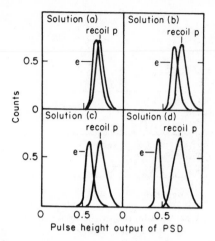

Fig. XV-14. Composition of four liquid scintillator solutions used by Jackson and Thomas (29) to measure pulse shape discrimination between neutrons of energy < 100 eV and γ rays. PBD—2-phenyl-5-(4-biphenylyl)-1,3,4-oxadiazole; IPBP—monoisopropyl biphenyl; EMB—enriched methyl borate (0.95 ^{10}B); POPOP—1,4-di-[2-(5-phenyloxazolyl)]-benzene; α-NPO—2-(1-naphthyl)-5-phenyloxazole; DPA—9,10-diphenylanthracene.

Solution	Composition by weight
(a)	PBD (1.4%) + IPBP (49.3%) + EMB (49.3%) + POPOP (20 mg/liter)
(b)	PBD (1.2%) + naphthalene (21.7%) + EMB (46.5%) + IPBP (30.6%)
(c)	α-NPO (1.2%) + naphthalene (21.7%) + EMB (46.5%) + IPBP (30.6%)
(d)	DPA (1.2%) + naphthalene (21.7%) + EMB (46.5%) + IPBP (30.6%)

Figure XV-14 shows four different liquid scintillator solutions and the pulse shape discriminator circuit response to the same γ-ray and neutron source. Some solution compositions are more efficient for either the production of triplet states or the prevention of their being deexcited by nonradiative processes (i.e., quenched). The solution used to obtain Fig. XV-14d is the best of those shown in the figure.

Other techniques have been used in attempts to obtain better discrimination at low light outputs. In one technique the neutrons were thermalized (reduced to very low energy by scattering) and finally trapped by ^{10}B which is present in the liquid scintillator (29–33). The ^{10}B then decays by production of an alpha particle of 2.3 or 2.8 MeV energy. The discriminator circuit then is used to differentiate between electron- and alpha-produced pulses in the liquid scintillator solution. The ^{10}B is incorporated into the solution as a methyl borate. See Chapter V for a discussion of neutron counting.

References

1. J. B. Birks, *in* "The Theory and Practice of Scintillation Counting," pp. 219–227 and references cited therein. Pergamon, Oxford, 1964.
2. R. Voltz, G. Laustriat, and A. Coche, *J. Phys. (Paris)* **29**, 159 (1968).
3. R. Voltz, H. DuPont, and G. Laustriat, *J. Phys. (Paris)* **29**, 297 (1968).
4. G. Laustriat, *in* "Organic Scintillators" (D. L. Horrocks, ed.), p. 127. Gordon & Breach, New York, 1968.
5. C. A. Parker and C. G. Hatchard, *Trans. Faraday Soc.* **57**, 1894 (1964).
6. C. A. Parker and C. G. Harchard, *Proc. Roy. Soc. Ser. A* **269**, 574 (1962).
7. T. A. King and R. Voltz, *Proc. Roy. Soc. Ser. A* **289**, 424 (1966).
8. W. L. Buck, *IRE Trans. Nucl. Sci.* **NS-7**, 11 (1960).
9. J. B. Birks, *IRE Trans. Nucl. Sci.* **NS-7**, 2 (1960).
10. P. E. Gibbons, D. C. Northrop, and O. Simpson, *Proc. Roy. Soc. London* **79**, 373 (1962).
11. F. D. Brooks, *Nucl. Instrum. Methods* **4**, 151 (1959).
12. F. D. Brooks, R. W. Pringle, and B. L. Funt, *IRE Trans. Nucl. Sci.* **NS-7**, 35 (1960).
13. L. M. Bollinger and G. E. Thomas, *Rev. Sci. Instrum.* **32**, 1044 (1961).
14. F. T. Kuchnir and F. J. Lynch, *IEEE Trans. Nucl. Sci.* **NS-15**, 107 (1968).
15. F. J. Lynch, *in* "Organic Scintillators" (D. L. Horrocks, ed.), p. 293. Gordon & Breach, New York, 1968.
16. M. L. Rousch, M. A. Wilson, and W. F. Hornyak, *Nucl. Instrum. Methods* **31**, 112 (1964).
17. W. Daehnick and R. Sherr, *Rev. Sci. Instrum.* **32**, 666 (1961).
18. T. K. Alexander and F. S. Goulding, *Nucl. Instrum. Methods* **13**, 244 (1961).
19. C. M. Ciaella and J. A. Devanney, *Nucl. Instrum. Methods* **60**, 269 (1968).
20. T. G. Miller, *Nucl. Instrum. Methods* **63**, 121 (1968).
21. G. W. McBeth, R. A. Winyard, and J. E. Lutkin, "Pulse Shape Discrimination with Organic Scintillators." Koch-Light Lab., Ltd., Colnbrook, Bucks, England, 1971.
22. H. W. Broek and C. E. Anderson, *Rev. Sci. Instrum.* **31**, 1063 (1960).
23. D. L. Horrocks, *Rev. Sci. Instrum.* **34**, 1035 (1963).
24. D. L. Horrocks, *Appl. Spectrosc.* **24**, 397 (1970).
25. R. B. Owens, *IRE Trans. Nucl. Sci.* **NS-9**, 285 (1962).
26. O. J. Hahn and R. C. Axtman, *Nucl. Instrum. Methods* **27**, 323 (1964).
27. D. W. Jones, *Nucl. Instrum. Methods* **62**, 19 (1968).
28. D. Metta, H. Diamond, R. F. Barnes, J. Milsted, J. Gray, Jr., D. J. Henderson, and C. M. Stevens, *J. Inorg. Nucl. Chem.* **27**, 33 (1965).
29. H. E. Jackson and G. E. Thomas, *Rev. Sci. Instrum.* **36**, 419 (1965).
30. L. M. Bollinger and G. E. Thomas, *Rev. Sci. Instrum.* **28**, 489 (1957).
31. C. O. Muehlhause and G. E. Thomas, *Phys. Rev.* **85**, 926 (1952).
32. L. M. Bollinger and G. E. Thomas, *Nucl. Instrum. Methods* **17**, 97 (1962).
33. G. E. Thomas, *Nucl. Instrum. Methods* **17**, 137 (1962).

CHAPTER XVI

FLOW CELL COUNTING

Many types of experiments lend themselves to continuous monitoring by a physical measurement. The measurement of the changing level of radio-activity in a sample has led to the development of flow cells that are sensitive to changes in the radioactivity levels. In recent years there has been an interest in the design and development of organic scintillator flow cells. These flow cells are of two types, continuous and discrete. They are also further distinguished by the relationship of the sample to the scintillator material, single phased and two phased.

Discrete Flow Cells

The discrete flow cell involves the collection of a certain volume of sample which is then introduced as a single sample into the counting system.

After counting, the sample is expelled automatically and the next sample is introduced. The rate of sample collection is often determined by the time required to count the sample to the desired statistics. If the specific activity of the sample is high, the counting time can be short and the collection time will be the time-determining factor.

Discrete flow cells can utilize several types of counting systems. Some of these are:

(a) premixing with liquid scintillation solution before introduction into the counting vial;

(b) direct introduction into a counting vial containing the scintillation detector in a second phase—plastic beads, anthracene crystals, etc.;

(c) Cerenkov counting of high-energy beta emitters in aqueous solutions.

Continuous Flow Cells

The continuous monitor of the source of sample by flowing the sample directly into the detector has been successfully applied to liquid scintillation counting systems. The flowing sample can be premixed with scintillation solution before introduction into the detector. The use of two-phase scintillation systems has also been demonstrated successfully for continuous flow systems.

Continuous flow cells have to be designed in such a manner that there is little mixing between successive sample volumes as they flow through the cell. Yet the volume being counted at any one time has to be large enough to get the desired counting statistics. Thus the active volume, flow rate, specific activity, desired statistical accuracy, and counting efficiency will all be important in the design of a continuous flow cell.

Many experiments have been performed using organic scintillators to monitor the effluents from gas chromatographs and ion exchange columns. The effluents flow directly through a packed column (plastic scintillator or anthracene crystals) which is placed in the counting chamber of a liquid scintillation counter. Figure XVI-1 shows some typical data obtained with an anthracene packed tube connected to the outlet of a gas chromatograph onto which had been loaded a series of ^{14}C-labeled fatty acids (1).

Two-Phased Flow Cells

The first flow cells developed were the heterogeneous-type cells, in which the sample and the scintillating material were in separate phases. The first

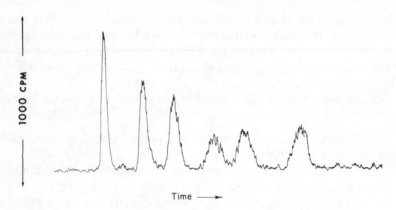

Fig. XVI-1.   Count rate of the effluent of a gas chromatograph used to separate a series of ^{14}C-labeled fatty acids.

such cells consisted of tubes of plastic scintillator through which an aqueous solution containing the sample flowed (2–4). The plastic tubing was formed into a coil to allow as large a volume as possible to be viewed by the multiplier phototubes. However, the tubing had to be of small diameter for two reasons. First, the small diameter prevented mixing and incomplete flushing of the sample from the cell. Second, to be efficient for counting the low-energy beta particles, the thickness of sample had to be small to allow the beta particles to reach the scintillator and produce excited molecules. However, a compromise had to be reached. If the diameter of tubing became too small, a backpressure could develop, which would delay the normal pumping of the solution.

The average beta-particle energy of ^{14}C decay is approximately 50 keV, and the E_{max} is 158 keV. The ranges of beta particles of these energies are 1.4×10^{-2} and 0.3 mm, respectively. Thus only those beta particles produced within 0.3 mm of the plastic walls will have a chance to excite the scintillator. And only those that are directed toward the wall (the beta particles are emitted isotropically) will have a chance to produce excited molecules.

The average beta energy of ^{3}H decay is approximately 6 keV and E_{max} is only 18 keV. The ranges of beta particles with these energies are only 5×10^{-4} and 5×10^{-3} mm, respectively. Thus the probability of excitation of the scintillator plastic walls is even less. Table XVI-1 shows the probability of measuring different energy beta particles in a flow cell with plastic scintillator walls and tubing of 4-mm inner diameter.

In cells of scintillator plastic tubing, efficiencies have been reported at 75% for ^{32}P, 50% for [22]Na, 6% for ^{14}C and ^{35}S, and less than 1% for [3]H. Similar counting efficiencies were obtained for a flow cell constructed of

Table XVI-1

The probability of measuring different energy beta particles in a flow cell[a]

Isotope	Energy (keV)	Range (mm)	Fraction of electrons with designated energy which will strike the wall of a 4-mm diameter tubing
^{3}H	6 (av.)	5×10^{-4}	<0.0010
	18 (E_{max})	5×10^{-3}	0.0025[b]
^{14}C	50 (av.)	1.4×10^{-2}	0.0064[b]
	158 (E_{max})	0.33	0.14[b]
^{32}P	570 (av.)	1.9	0.50[b]
	1710 (E_{max})	8.1	1.00[c]

[a] Flow cell has plastic scintillator walls.

[b] If the range is less than the tube diameter, it is necessary to include a factor of $\frac{1}{2}$ due to equal probability of beta particle emitted toward or away from the plastic tube walls.

[c] With the very high energy it is possible that the beta particles will pass through the plastic walls of the tubing.

two parallel sheets of plastic scintillator separated by a sheet of nonscintillating plastic through which holes had been drilled for the flow of the sample (5). These flow cells were inadequate for the measurement of low-energy beta emitters, such as ^{3}H.

Packed flow cells were demonstrated to have much higher counting efficiencies than the scintillating wall type (6–8). Several different types of organic crystal scintillators were investigated. Anthracene crystals, because of their high scintillation efficiency ($\sim 4\%$), were found to have the highest counting efficiency (9–11). Proper choice of crystal size, diameter of cell, cell material, and other parameters (12) have given ^{14}C and ^{3}H efficiencies up to 55 and 2%, respectively.

Some major problems occur in flow cells using anthracene crystals. Since the crystals are organic, care is necessary in the choice of the solvent that is used to flush the sample through the cell, since certain solvents will dissolve the anthracene. Some samples have a tendency to absorb or react with the anthracene, leading to memory problems. In some cases the anthracene crystals have tended to oxidize on the surface, thus reducing the scintillation efficiency and acting as a light absorber for the scintillations (i.e., color quenching).

Different geometry cells packed with anthracene crystals have been used. Spiral tubes (10), straight tubes (12), and others (13, 14) have been used successfully. One major problem is the packing and repacking of the cells. Once the anthracene crystals are no longer useful, it is necessary either to be able to shake out the crystals or dissolve them with solvents. In some cells the repacking has been accomplished by using a suspension of the anthracene crystals which is forced into the cell. After settling the crystals are maintained in the cell by plugging the ends with quartz wool, so the crystals will not float out.

Plastic beads have been used in place of anthracene crystals, but the counting efficiency is much less. However, plastic beads are more efficient than just plastic walls. Filaments of plastic scintillators have been packed in bundles inside a glass tube, and the sample is drawn into the voids by capillary action. Counting efficiencies up to 35% for ^{14}C and 0.7% for ^{3}H were observed (6–8). However, because of the capillary action it is very difficult to flush out the cell and almost impossible to use it with a continuous flowing sample.

Recently Osborne (15) reported a flow cell made of many thin sheets of plastic scintillator placed very close and parallel. The sample flowed through the cell between the sheets of plastic. The detector was capable of measuring tritium levels down to 1 μCi/liter of water.

Single-Phased Flow Cells

In the single-phased flow cell the sample and liquid scintillator solutions are mixed prior to introduction into the counting chamber. This can be done in a continuous or in a discrete manner. Immediately, it is seen that certain requirements are essential in order that the continuous homogeneous flow cell operate properly. Some of these are:

1. complete solubility of the sample in the scintillator solution. In this type of system emulsion solutions are often used, and it should be kept in mind that emulsions are not true homogeneous solutions;

2. complete mixing of sample and scintillator solution before they reach the counting chamber;

3. small chamber for mixing to reduce memory and increase resolution of different levels of radioactivity;

4. flow rates which match column separation requirements and counting times necessary to obtain statistically reliable measurement;

5. freedom from changing quench, chemiluminescence, etc., which are difficult (although not impossible) to monitor in the continually changing system.

Flow Cell Application to Measuring Tritium in Water

Recently, single-phased continuous flow cell techniques have been applied to monitoring tritium levels in water (16) with the goal of measuring tritium levels in nuclear power plant coolant systems. Successful operation was dependent on the elimination of several serious problems: light piping causing a high background, proper premixing of water and scintillator solutions, large enough volume of sample to get statistically reliable data, etc. Figure XVI-2 shows a block diagram of the sample mixing and flow cell apparatus.

A unique feature of this flow cell is the introduction of light traps at the inlet and outlet of the counting chamber. Normally ambient light is piped down the sample transport tubes and causes exposure to the multiplier phototubes. This will cause a high background and make it impossible to measure low levels of radioactivity. The light traps consisted of a glass coil with an irregular innersurface and black paint on the outside of the tubing transporting the sample to the counting cell. The irregular inner surface prevented direct piping of light into the counting cell.

The system response was checked by injection of discrete, small volumes of tritiated water. The results are shown in Fig. XVI-3. The sample count rate rose to a maximum as the ^{3}H-water passed through the cell and returned to the normal background rate. Injections were made at 7-min intervals. The data show essentially no memory or leakage of the ^{3}H between injections.

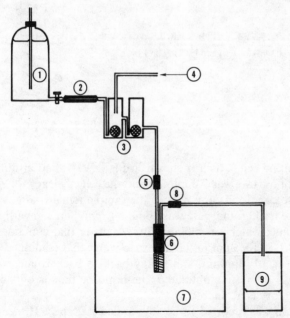

Fig. XVI-2. Continuous flow cell and liquid scintillation system. (1) Constant-head reservoir; (2) glass capillary restrictor; (3) mixing chamber; (4) sample inlet; (5) light trap; (6) probe assembly; (7) liquid scintillator counter; (8) light trap; (9) waste container.

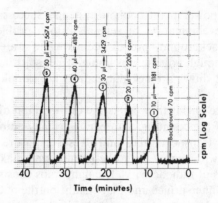

Fig. XVI-3. Flow cell monitor showing response to discrete injections of THO into water stream. Analog record of five successive injections (scintillator flow rate 5.4 ml/min).

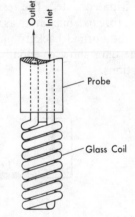

Fig. XVI-4.   Flow cell (effective volume
4.3 ml) as designed by Ting and Litle (16).

The counting cell, which is shown in Fig. XVI-4, had a total volume of 4.3 ml. The flow rate was 7.0 ml/min and the cell efficiency for ^3H was 32%. Under these conditions any part of the sample remains almost 3.9 min in the cell. The background rate was about 50 cpm. Starting with background water, the inlet water was changed first to water that contained tritium at a level of 8.1×10^3 dpm/ml, and next to water that contained tritium at a level of 6.5×10^4 dpm/ml, and finally back to background water. Figure XVI-5 shows the results obtained. The response time is only about 3 min and the memory is negligible.

The system was reported to be able to detect 25 pCi ^3H/ml water. However, there is still the problem of quenching that can be produced by impurities in the water. More important is the possibility of variable quench.

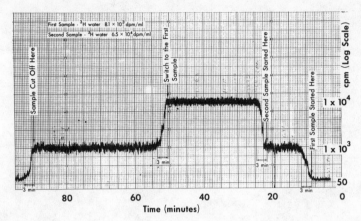

Fig. XVI-5. Continuous monitoring response to change of activity. First sample—^{3}H-water, 8.1 × 10^3 dpm/ml; second sample—^{3}H-water, 6.5 × 10^4 dpm/ml.

References

1. A. Karmen, Packard Tech. Bull. No. 14. Packard Instrum. Co., Inc., Downers Grove, Illinois, 1965.
2. J. W. Rudeman, U.S. Patent 2,961,541, November 22, 1960.
3. E. Schram and R. Lombaert, *Anal. Chim. Acta* **17,** 417 (1957).
4. E. Schram and R. Lombaert, *Biochem. J.* **66,** 21 (1957).
5. E. Shram, *in* "The Current Status of Liquid Scintillation Counting" (E. D. Bransome, Jr., ed.), p. 95. Grune & Stratton, New York, 1960.
6. D. Steinberg, *Nature (London)* **182,** 740 (1958).
7. D. Steinberg, *Nature (London)* **183,** 1253 (1959).
8. D. Steinberg, *Anal. Biochem.* **1,** 23 (1960).
9. E. Schram and R. Lombaert, *Arch. Int. Physiol. Biochim.* **68,** 845 (1960).
10. E. Schram and R. Lombaert, *Anal. Biochem.* **3,** 68 (1962).
11. E. Rapkin and L. E. Packard, *Proc. Conf. Organic Scintillation Detectors, Univ. of New Mexico, 1960* (G. H. Daub, F. N. Hayes, and E. Sullivan, eds.), TID-7612, p. 216. U. S. At. Energy Commission, Washington, D. C., 1961.
12. E. Rapkin and J. Gibbs, *Nature (London)* **194,** 34 (1962).
13. K. A. Piez, *Anal. Biochem.* **4,** 444 (1962).
14. D. H. Elwyn, *Advan. Tracer Methodol.* **2,** 115 (1965).
15. R. V. Osborne, *in* "Tritium" (A. A. Moghissi and M. W. Carter, eds.), p. 496. Messenger Graphics, Las Vegas, Nevada, 1973.
16. P. Ting and R. L. Litle, *in* "Tritium" (A. A. Moghissi and M. W. Carter, eds.), p. 170. Messenger Graphics, Las Vegas, Nevada, 1973.

CHAPTER XVII

LARGE-VOLUME COUNTERS

Most applications which involve large volumes of liquid scintillator solutions are for measurement of activity levels or detection of special particles which are external to the detector. Some recent applications are now being pursued which involve volumes of up to a liter with internal samples, i.e., large volumes of water for detection of very low levels of tritium.

Because of the large quantities of materials used in the large-volume detectors, the purity of all the components is very critical. The photons have to travel long distances to reach the multiplier phototubes, which are placed in the walls of the cavity holding the scintillator solution. Any minute traces of impurities which absorb even a small fraction of the photons per unit distance will cause a greater decrease in response due to the long path length. The intensity I will decrease according to

$$\log I/I_0 = \epsilon l c.$$

If the extinction coefficient ϵ and the concentration c remain the same, an increase in the path length l will decrease the intensity by the calculated

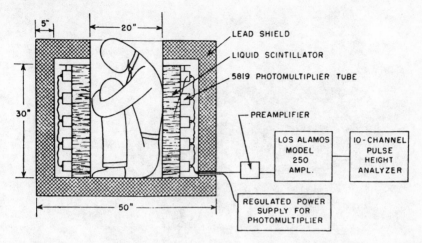

Fig. XVII-1. The first human counter using liquid scintillators at Los Alamos in 1953 (1).

amount. The use of secondary solutes (wavelength shifters) is a necessity to reduce the value of ϵ greatly for the wavelength of the emitted photons.

Whole-Body Counters

A major application of large-volume liquid scintillation counters is the measurement of radioactivity in animals and humans. These counters come in various arrangements, some of which are shown in Figs. XVII-1–XVII-4.

Fig. XVII-2. Arrangement of three $20 \times 10 \times 6\frac{1}{2}$-in. plastic scintillator units with respect to seated patient (2).

Fig. XVII-3.   Artist's sketch of Humco II whole-body counter at Los Alamos (3).

They have been designed with the desire to provide as much comfort as possible to the patient while allowing for maximum counting efficiency and information concerning which different radionuclides are present.

The design of the whole-body liquid scintillation counters has several important factors for consideration. The stopping or scattering properties of the scintillator solution will determine the counting efficiency and energy resolution that can theoretically be obtained. The volume and thickness of the scintillator solution will determine the probability that a γ ray will undergo single Compton scattering, multiple Compton scattering, and/or total absorption. The higher the energy of the γ ray to be measured, the greater is the thickness of the scintillator solution required for multiple scattering or total absorption. Since the body contains ^{40}K which emits γ rays of 1.46 MeV, most counters should have at least a thickness of scintillator solution which will give a favorable number of total absorptions of these γ rays. However, the thickness will be limited by several factors:

(a) the bulk and purity of materials required to fill a very large cavity, cost, availability, etc.;

(b) light collection efficiency, the number of multiplier phototubes required to view any practical area of the surface of the counter cavity and the optical reflector properties;

(c) size of counter cavity and construction costs;

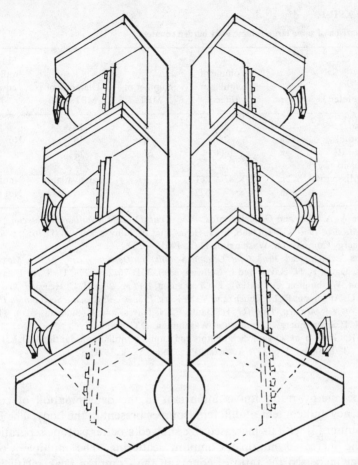

Fig. XVII-4. Suggested arrangement for a six-tank liquid scintillation counter for children and adults in use at the University of Missouri (4).

(d) background increases with increased size of detector;

(e) necessary shielding to reduce background, size and bulk of shielding, and/or anticoincidence detectors.

Some properties of several whole-body liquid scintillation counters are given in Table XVII-1. In 2π counters there may be errors due to the geometry of the source of the radioactivity in the body. To a great extent this effect is reduced, although not eliminated, in 4π counters. However, in all such counters masking by different parts of the body is important in determining the body burden.

Table XVII-1

Comparison of some large-volume body burden counters

Counter	Type	Volume of scintillator (liters)	Number of MPTs	Diameter of MPTs (in.)	Fraction of wall surface covered by MPT
Genco[a]	2π	300	6	14½	0.18
Landstuhl[b]	2π	280	6	16	Not available
Humco II[c]	4π	1600	24	14½	0.24
University of Missouri[d]	2π	$3 \times 87 = 261$	3	Not available	Not available
	4π	$6 \times 87 = 522$	6	Not available	Not available

[a]See A. A. Pfan and G. Kallistratos, *Proc. Conf. Organic Scintillation Detectors, Univ. of New Mexico, 1960* (G. H. Daub, F. N. Hayes, and E. Sullivan, eds.), TID-7612, p. 293. U.S. At. Energy Commission, Washington, D.C., 1961.

[b]See C. O. Onstead, *Proc. Conf. Organic Scintillation Detectors, Univ. of New Mexico, 1960* (G. H. Daub, F. N. Hayes, and E. Sullivan, eds.), TID-7612, p. 278. U. S. At. Energy Commission, Washington, D. C., 1961; E. Oberhausen, *Ibid.*, p. 286; H. C. Heinrich, *Ibid.*, p. 312.

[c]E. C. Anderson, R. L. Schuch, and V. N. Kerr, *Proc. Conf. Organic Scintillation Detectors, Univ. of New Mexico, 1960* (G. H. Daub, F. N. Hayes, and E. Sullivan, eds.), TID-7612, p. 344. U.S. At. Energy Commission, Washington, D.C., 1961.

[d]E. R. Graham, *in* "Organic Scintillators and Liquid Scintillation Counting" (D. L. Horrocks and C. T. Peng, eds.), p. 137. Academic Press, New York, 1971.

The energy resolution is important in the determination of counting efficiency and content of different isotopes present in the body. The Humco II counter (3) with its increased size showed a better isotope separation due to the increased multiple Compton scattering. The multiple Compton scatter increased the ratio of counts in the Compton peak relative to the lower energy Compton continuum. Figure XVII-5 shows typical pulse height spectra obtained with the Humco II counter. It is possible to set three separate counting channels to measure ^{137}Cs, ^{65}Zn, and ^{40}K in a single patient. The observed counts in each channel would have to be corrected for the contribution of each isotope.

The energy calibration is necessary to ensure the proper relation between selected counting channels and isotope γ-ray energy. Critical in the calibration is the knowledge of the type of interaction produced in the liquid scintillator, i.e., single Compton scatter, multiple Compton scatter, or total absorption. Figure XVII-6 shows the energy calibration measured for the Humco II based on a double Compton scatter of the γ ray. The response is linear with the deposited energy.

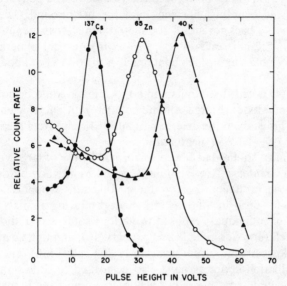

Fig. XVII-5. Typical pulse height spectra obtained with the Humco II whole-body counter (3).

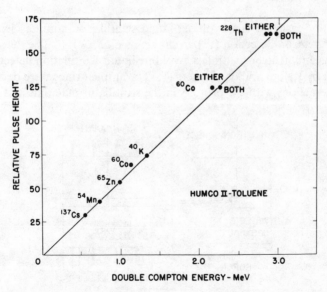

Fig. XVII-6. Energy calibration of the Humco II whole-body counter (3).

The detection limits of these large-volume whole-body counters will depend on the counting efficiency and the background. The Humco II counter (3) has a background of about 300 counts/sec in a counting channel which has a ^{40}K efficiency of 2%, i.e., a 70-kg-weight male adult should have 133 g of potassium which contains 15.8 mg of ^{40}K, or about 424 γ rays/sec.

Graham (4) noted that adults and large animals could be counted in the University of Missouri large-volume counter with ^{40}K efficiencies as high as 47.3%. This enabled the determination of ^{40}K levels to a mean coefficient of variation of $\pm 1.7\%$ in just 6 min.

Many clinical studies have been performed by use of large-volume liquid scintillation counters: ^{60}Co–vitamin B_{12} uptake and body burden (5); presence of fission products from fallout in foodstuffs, animals, and humans (6); lean-to-fat ratio in humans and meat-producing animals (7, 8); contamination of individuals exposed to various sources of radioactivity (9); red-blood cell turnover in leukemic patients (10); and uptake and retention of such biologically important trancers as ^{59}Fe and ^{131}I (8). The whole-body liquid scintillation counters have also been used to follow the biological half-life of gamma-emitting radionuclides in animals and humans.

Other Applications

Specially designed large-volume liquid scintillation counters have been used for neutron detection (11) as shown in Fig. XVII-7. The detection of neutrinos and neutrons with large-volume liquid scintillation detectors was reported by Reines and co-workers (1). The antineutrinos were detected by their interactions with the protons in the scintillator materials and efficient

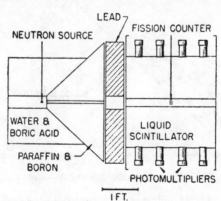

Fig. XVII-7. The Los Alamos large-volume liquid scintillation counter for detecting neutrons (14).

detection of the resulting positron (annihilation quanta) and neutron (capture γ rays). They have also been used to detect cosmic ray particles, time and flight measurement of neutrons, and scintillation track imaging (2).

Solvent Requirements

The large volumes not only dictate the need for very pure solvents, but also, for safety reasons, require those that have a low flash point. The commercially available solvent TS-28M (Shell Oil Co.) has a flash point of 49°C and has been investigated at Los Alamos (3). The TS-28M is very difficult to obtain in the needed purity. Trimethylbenzene is an ideal solvent for this application but it is hard to obtain in the high purity required and is very expensive. Most large-volume counters use toluene as the solvent, flash point 45°C.

References

1. F. Reines, *in* "Liquid Scintillation Counting" (C. G. Bell, Jr., and F. N. Hayes, eds.), p. 246. Pergamon, Oxford, 1958.
2. P. R. J. Burch, *Proc. Conf. Organic Scintillation Detectors, Univ. of New Mexico, 1960* (G. H. Daub, F. N. Hayes, and E. Sullivan, eds.), TID-7612, p. 329. U. S. At. Energy Commission, Washington, D. C., 1961.
3. E. C. Anderson, R. L. Schuch, and V. N. Kerr, *Proc. Conf. Organic Scintillation Detectors, Univ. of New Mexico, 1960* (G. H. Daub, F. N. Hayes, and E. Sullivan, eds.), TID-7612, p. 344. U. S. At. Energy Commission, Washington, D. C., 1961.
4. E. R. Graham, *in* "Organic Scintillators and Liquid Scintillation Counting" (D. L. Horrocks and C. T. Peng, eds.), p. 137. Academic Press, New York, 1971.
5. H. C. Heinrich, *Proc. Conf. Organic Scintillation Detectors, Univ. of New Mexico, 1960* (G. H. Daub, F. N. Hayes, and E. Sullivan, eds.), TID-7612, p. 312. U. S. At. Energy Commission, Washington, D. C., 1961.
6. E. C. Anderson, *in* "Liquid Scintillation Counting" (C. G. Bell, Jr. and F. N. Hayes, eds.), p. 211. Pergamon, Oxford, 1958.
7. K. T. Woodward, T. T. Trujillo, R. L. Schuch, and E. C. Anderson, *Nature (London)* **178,** 97 (1956).
8. W. H. Langham, *in* "Liquid Scintillation Counting" (C. G. Bell, Jr. and F. N. Hayes, eds.), p. 135. Pergamon, Oxford, 1958.
9. C. R. Richmond and W. H. Langham, *Proc. Health Phys. Soc. Ann. Meeting, 1st, Univ. of Michigan, Ann Arbor, 1956.* Univ. of Michigan, Ann Arbor, 1957.
10. C. C. Lushbaugh, Unpublished results. Los Alamos Sci. Lab., Los Alamos, New Mexico, 1957.
11. F. B. Harrison, *in* "Liquid Scintillation Counting" (C. G. Bell, Jr. and F. N. Hayes, eds.), p. 150. Pergamon, Oxford, 1958.
12. R. Reines, *Proc. Conf. Organic Scintillation Detectors, Univ. of New Mexico, 1960* (G. H. Daub, F. N. Hayes, and E. Sullivan, eds.), TID-7612, p. 395. U. S. At. Energy Commission, Washington, D. C., 1961.

CHAPTER XVIII

STATISTICAL CONSIDERATIONS

The assumption can be made that the results of an experiment subject to random error, if repeated sufficiently often, will "average out" to the "true" result of an ideal experiment not subject to random errors. Most of the actual results will be close to the ideal value, fewer of them somewhat less close, and a very small number quite far away. This observation leads to the concept of a *distribution function* for an experimental measurement subject to random errors. The distribution function is one which describes the fraction of times in a large number of measurements that a measurement falls between two specified values. The phrase "fraction of times in a large number of measurements" is an approximate definition of *probability*. The rigorous definition is the limit of that fraction as the number of measurements increases without limit.

Distributions can, of course, arise from sources other than errors in the measurement of some simple precise quantity. If a property (i.e., weight) of a number of supposedly identical objects is measured, some variation will be observed, part of which may be due to random errors in the measurement process (weighing process) and part to actual differences between the objects themselves. In this case the "averaging out" process carried out over a number of measurements refers not to the "true" value (weight) of any actual object but to the kind of idealized or typical property (weight) one would expect it to have.

Table XVIII-1

Data obtained with a counter measuring a steady source

Measurement	Counts	Δ^a	Δ^2
1	203	−0.4	0.16
2	194	−9.4	88.36
3	201	−2.4	5.76
4	217	13.6	184.96
5	195	−8.4	70.56
6	189	−14.4	207.36
7	210	6.6	43.56
8	207	3.6	12.96
9	230	26.6	707.56
10	188	−15.4	237.16
Total	2034	0	1558.40

$^a\Delta$ is the fluctuation from the average.

Table XVIII-1 gives a set of data obtained with a counter measuring a "steady" source. Since the number of counts recorded per minute varies, an experimenter will want to know which minute count was the most accurate, i.e., most likely to be the true counting rate, the best estimate of which is the arithmetic mean (the average value). What we are trying to do is to estimate from a finite number of observations the results of an essentially infinite number of observations. In particular, we wish to estimate the average value and the distribution of the observed values about that average.

Average Value

The average value of a sampling of data is defined as

$$\overline{X} = \frac{1}{N_0} \sum_{i=1}^{N_0} X_i. \tag{1}$$

For the example in Table XVIII-1, $\overline{X} = 2034/10 = 203.4$. This average value is the best estimate possible from the data of the true average $\overline{X}_t$.

Standard Deviation

The distribution of the observed results about $\overline{X}_t$ is a measure of the precision of the data and can be considered as all of the moments of the

$$\frac{1}{N_0} \sum_{i=1}^{N_0} (X_i - \overline{X}_t)^n \tag{2}$$

distribution for $n = 1, 2, 3, \ldots$. The first moment $(n = 1)$ will always be equal to zero because of the definition of $\overline{X}_t$. The other odd moments (i.e., n an odd number) will vanish only if the distribution is symmetrical about $\overline{X}_t$, in which case $\overline{X}_t$ is the most probable value of X. Usually, only the second moment $(n = 2)$ is used in practice and is called the variance, denoted by σ_x^2. The square root of the variance is called the standard deviation σ_x.

The variance is important because of its use in evaluating the so-called normal distribution law, which is expected to describe the distribution of experimental results with random errors:

$$P(x)\, dx = \frac{1}{(2\pi\sigma_x^2)^{1/2}} \exp\left[\frac{-(X - \overline{X}_t)^2}{2\sigma_x^2}\right] dx \tag{3}$$

where $P(x)\, dx$ is the probability of observing a value of x in the interval x to $x + dx$.

For the example given in Table XVIII-1, a finite number of observations are available, so we do not know $\overline{X}_t$, only an estimate of it, $\overline{X}$. Under these circumstances, the best possible estimate of the variance is

$$\sigma_x^2 = \frac{1}{(N_0 - 1)} \sum_{i=1}^{N_0} (X_i - \overline{X})^2. \tag{4}$$

For the data in Table XVIII-1,

$$\sigma_x^2 = \tfrac{1}{9}(1558.4) = 173.2$$

$$\sigma_x = 13.2.$$

The appearance in Eq. (4) of $(N_0 - 1)$ rather than N_0 as in Eq. (2) is a consequence of estimating the unknown quantity $\overline{X}_t$ from N_0 observations. This estimation uses up one of the observations and leaves only $(N_0 - 1)$ independent quantities for the estimation of the variance.

Precision of the Average Value

The precision of our estimation is not to be confused with the precision of the data, although the two quantities are related. Two things are of concern:

1. the distribution of the values of $\overline{X}$ given by Eq. (1) from many sets of experiments, each with a finite N_0, and

2. the distributions of the quantities σ_x^2 obtained from the same sets of observations by Eq. (4).

A measure of the reliability of $\overline{X}$ is the variance of the mean, which is estimated by the variance of the set of observations divided by N_0:

$$\sigma_{\overline{x}}^2 = \frac{\sigma_x^2}{N_0} = \frac{1}{N_0(N_0 - 1)} \sum_{i=1}^{N_0} (X_i - \overline{X})^2. \tag{5}$$

The quantity $\sigma_{\overline{x}}^2$ is the best estimate of the second moment of the distribution of the average values that would be found from an infinite number of sets of experiments, each containing N_0 observations. For the data in Table XVIII-1,

$$\sigma_{\overline{x}}^2 = 173.2/10, \qquad \sigma_{\overline{x}} = (17.32)^{1/2} = 4.2, \qquad \overline{X} = 203.4 \pm 4.2.$$

The significance of the smaller variance for the average can be seen from the distribution of a large number of cases. The distribution of the averages would have a smaller spread.

An illustration of these two types of distributions is given in Table XVIII-2. Ten aliquots from a stock solution of ^{241}Am were dissolved in 10 different liquid scintillator solutions. Each of the samples was counted for 10 min, and one of the samples was counted 10 times for 10 min each. Because the 10 different samples have errors other than the counting rate fluctuation (namely pipet errors), the fluctuation from average Δ is greater than for the single sample.

Binomial Distribution

The binomial distribution law, which can be derived by application of the addition and multiplication theorems, treats one fairly general case of compounding probabilities.

Addition. For two mutually exclusive events with probabilities p_1 and p_2 the probability of one or the other occurring is $p_1 + p_2$, or in more general terms,

$$\sum_{i=1}^{j} p_i.$$

Table XVIII-2a

Counting one sample ten times

Measurement	Counts/10 min	Δ^a	Δ^2
1	19,750	124	15,376
2	19,875	249	62,001
3	19,443	-183	33,489
4	19,499	-127	16,129
5	19,793	167	27,889
6	19,558	-68	4624
7	19,675	49	2401
8	19,325	-299	89,401
9	19,761	135	18,225
10	19,579	-47	2209
Total	196,258	0	271,744

$$\overline{X} = 19,626$$

Variance of the set: $\sigma^2 = [1/(N_0 - 1)] \sum \Delta^2 = 271,744/9 = 30,194$

$$\sigma = 173.8, \quad \sigma/\overline{X} = 0.0089 \quad \text{or} \quad 0.89\%$$

Variance of the average: $\sigma_{\text{Av}}^2 = [1/N_0(N_0 - 1)] \sum \Delta^2 = 30,194/10 = 3019$

$$\sigma_{\text{Av}} = 54.9 \quad [0.0028 \quad \text{or} \quad 0.28\% = \sigma_{\text{Av}}/\overline{X}]$$

$$\overline{X} = 19,626 \pm 55$$

$^a\Delta$ is the fluctuation from the average.

Example. When a coin is tossed, the probability of either heads or tails is $1/2 + 1/2 = 1$. When a die is rolled, the probability of either 3 or 5 is $1/6 + 1/6 = 2/6 = 1/3$.

Multiplication. The probability of first one event occurring followed by a second event is the product of the two probabilities:

$$p_i \times p_j.$$

Example. When a coin is tossed twice, the probability of heads twice is $1/2 \times 1/2 = 1/4$. When a die is rolled twice, the probability of two ones is $1/6 \times 1/6 = 1/36$.

Given a large set of objects in which the probability of occurrence of an object of a particular kind w is p, then if n objects are withdrawn from the set, the probability $W(r)$ that exactly r of the objects are of the kind w is given by

$$W(r) = \frac{n!}{(n - r)!r!} p^r(1 - p)^{n-r}, \tag{6}$$

Table XVIII-2b

Counting of ten samples

Sample No.	Counts/10 min	Δ^a	Δ^2
1	17,557	−744	553,536
2	17,793	−508	258,064
3	17,945	−356	126,736
4	18,078	−223	49,729
5	18,108	−193	37,249
6	18,498	197	38,809
7	18,421	120	14,400
8	17,976	−325	105,625
9	19,310	1009	1,018,081
10	19,322	1021	1,042,441
Total	183,008	−2	3,244,670

$$\bar{X} = 18,301$$

$$\sigma^2 = [1/(N_0 - 1)] \sum \Delta^2 = 3,244,670/9 = 360,519$$

$$\sigma = 600.5, \quad \sigma/\bar{X} = 0.0328 \quad \text{or} \quad 3.28\%$$

$$\sigma^2_{\text{Av}} = [1/N_0(N_0 - 1)] \sum \Delta^2 = 360,519/10 = 36,051.9$$

$$\sigma_{\text{Av}} = 189.9 \quad [0.0104 \quad \text{or} \quad 1.04\% = \sigma_{\text{Av}}/\bar{X}]$$

$$\bar{X} = 18,301 \pm 190.$$

$^a\Delta$ is the fluctuation from the average.

and the binomial distribution is normalized:

$$\sum_{r=0}^{n} W(r) = 1. \tag{7}$$

Binomial Distribution for Radioactive Decay

The binomial distribution law can be applied to find the probability $W(m)$ of obtaining just m disintegrations in time t from N_0 original atoms, where N_0 is the number n of objects chosen for observation, and m is the number r that disintegrates in the time t:

$$W(m) = \frac{N_0!}{(N_0 - m)! \, m!} \, p^m (1 - p)^{N_0 - m}. \tag{8}$$

$1 - p$, the probability of an atom not decaying in the time t, is the ratio of the number N that survive the time interval t to the initial number N_0:

$$1 - p = N/N_0 = e^{-\lambda t}.$$

Then p is

$$p = 1 - (1 - p) = 1 - e^{-\lambda t},$$

and the binomial equation becomes

$$W(m) = \frac{N_0!}{(N_0 - m)!m!}(1 - e^{-\lambda t})^m (e^{-\lambda t})^{N_0 - m}. \tag{9}$$

Example. Experimenters A and B each place six atoms of a radioactive isotope of known half-life into a 4π counter (100% detection efficiency).

(a) Experimenter A presets his counting interval to equal the half-life of the isotope. What is the probability that he will observe $0, 1, \ldots, 6$ disintegrations, or what is $W(0), W(1), \ldots, W(6)$? Since in this case $N_0 = 6$,

$$W(m) = \frac{6!}{(6 - m)!m!}\left(\frac{1}{2}\right)^m\left(\frac{1}{2}\right)^{6-m} = \frac{1}{64}\frac{6!}{(6 - m)!m!}$$

and

$$W(0) = W(6) = \frac{1}{64}\frac{6!}{6!0!} = \frac{1}{64}$$

$$W(1) = W(5) = \frac{1}{64}\frac{6!}{5!1!} = \frac{6}{64}$$

$$W(2) = W(4) = \frac{1}{64}\frac{6!}{4!2!} = \frac{15}{64}$$

$$W(3) = \frac{1}{64}\frac{6!}{3!3!} = \frac{20}{64}$$

$$\Sigma\, W(m) = 2\left(\frac{1}{64}\right) + 2\left(\frac{6}{64}\right) + 2\left(\frac{15}{64}\right) + \frac{20}{64} = \frac{64}{64} = 1.$$

(b) Experimenter B wishes to adjust the counting interval to have a probability of $1/2$ of observing six disintegrations, i.e., $W(6) = 0.5$. How long should the time interval be?

$$W(6) = 0.5 = \frac{6!}{6!0!}(1 - e^{-\lambda t})^6 (e^{-\lambda t})^0 = (1 - e^{-\lambda t})^6.$$

Thus,

$$1 - e^{-\lambda t} = \left(\frac{1}{2}\right)^{1/6}, \quad \text{or} \quad 1 - \left(\frac{1}{2}\right)^{1/6} = e^{-\lambda t},$$

so

$$\ln\left[1 - \left(\frac{1}{2}\right)^{1/6}\right] = -\lambda t.$$

Therefore,

$$t = \frac{\ln[1 - (1/2)^{1/6}]}{\ln 2} t_{1/2} = 3.19 t_{1/2}.$$

The probability of any one of the N_0 atoms disintegrating in the time dt is $dN = N_0 \lambda \, dt$.

Average Disintegration Rate

The binomial law can be applied to radioactive disintegrations to calculate the average value of a set of numbers obeying the binomial distribution law. Using Eq. (6) and replacing $1 - p$ by q,

$$W(r) = \frac{n!}{(n-r)!r!} p^r q^{n-r}. \tag{10}$$

From

$$\overline{X} = \frac{1}{N_0} \sum n_i x_i = \sum p_i x_i$$

or more generally

$$\overline{f(x)} = \sum p_i f(x_i),$$

the average value of r is

$$\bar{r} = \sum_{r=0}^{n} r W(r) = \sum_{r=0}^{n} r \frac{n!}{(n-r)!r!} p^r q^{n-r}.$$

To evaluate this summation, consider the binomial expansion

$$(px + q)^n = \sum_{r=0}^{n} \frac{n!}{(n-r)!r!} p^r x^r q^{n-r} = \sum_{r=0}^{n} x^r W(r).$$

Differentiating with respect to x,

$$np(px + q)^{n-1} = \sum_{r=0}^{n} r x^{r-1} W(r). \tag{11}$$

Letting $x = 1$ and $q = 1 - p$,

$$np(p + 1 - p)^{n-1} = \sum_{r=0}^{n} r W(r), \qquad np = \bar{r},$$

which means that the average number $\bar{r}$ of the n objects which are of the kind w is just n times the probability for any given one of the objects to be of the kind w.

Interpreting these results for radioactive disintegration, we will set $n = N_0$ and $p = 1 - e^{-\lambda t}$. Then the average number M of atoms disintegrating in the time t is

$$M = N_0(1 - e^{-\lambda t}).$$

For small values of λt, i.e., for counting times very short compared to the half-life, the approximation

$$e^{-\lambda t} = 1 - \lambda t,$$

so that

$$M = N_0 \lambda t.$$

The disintegration rate R is then

$$R = M/t = \lambda N_0 \qquad (-dN/dt = \lambda N).$$

Expected Standard Deviation

Differentiating Eq. (11) again with respect to x gives

$$n(n-1)p^2(px+q)^{n-2} = \sum_{r=0}^{n} r(r-1)x^{r-2}W(r).$$

Again, letting $x = 1$ and using $p + q = 1$,

$$n(n-1)p^2 = \sum_{r=0}^{n} r(r-1)W(r) = \sum_{r=0}^{n} r^2 W(r) - \sum_{r=0}^{n} rW(r)$$

$$= \overline{r^2} - \bar{r}.$$

The variance is also given by

$$\sigma_r{}^2 = \overline{r^2} - \bar{r}^2.$$

$$= n(n-1)p^2 + \bar{r} - \bar{r}^2,$$

and using $\bar{r} = np$,

$$\sigma_r{}^2 = n^2 p^2 - np^2 + np - n^2 p^2 = np(1-p)$$

$$= npq$$

$$\sigma_r = (npq)^{1/2}.$$

For radioactive disintegrations

$$\sigma = (N_0(1 - e^{-\lambda t})e^{-\lambda t})^{1/2} = (Me^{-\lambda t})^{1/2}.$$

In counting practice λt is usually small:

$$e^{-\lambda t} \rightarrow 1.$$

That is, $\sigma = \sqrt{M}$ or, as we will see later, $\sigma = K\sqrt{M}$. This same simple relationship can also be derived for Poisson and Gaussian distributions. Thus a single observation of the radioactive disintegration rate gives both an estimate of the mean and an estimate of the variance of the distribution.

The statistical accuracy (fractional error of a counting rate) depends only on the total number of counts taken. Thus the counting time necessary to obtain a certain accuracy varies inversely as the counting rate, since total counts = (counting rate)(time of count). Because the laws of probability apply, one can never be entirely certain of one's results. The measured rate may not be the true rate, the value determined with an infinite number of counts, but it is possible to estimate the likelihood that the true rate will lie within some specified limits centering around the determined values.

For every set of limits there is a different probability that the result is correct to within those limits, and this probability is greater the broader the limits. For example, if the counting rate is determined to be 200 cpm, it is more likely that the true rate lies between 190 and 210 than between 199 and 201. The limits are often expressed as the fractional error, i.e., the fraction of the rate. Thus ± 10 would be $10/200$ or 0.05 fractional error, and ± 1 would be $1/200$ or 0.005 fractional error.

If ρ designates the fractional error and W the probability of making an actual error less than ρ, then $\rho = K/\sqrt{M}$, and since $\sigma = K\sqrt{M}$,

$$\rho = \frac{K\sqrt{M}}{M} = \frac{\sigma}{M},$$

where M is the total number of counts and K is a constant depending on W (see Table XVIII-3).

Example. For a counting rate of approximately 900 cpm and a counting time of 9 min, $M = 8100$. The standard error (68.3%) is given by:

$$\frac{1.0}{(8100)^{1/2}} = \frac{1}{90} = 0.011$$

$$(900)(0.011) = 10 \qquad (900 \pm 10).$$

Table XVIII-3

Values of probability constant K

K	Probability that an actual fractional error is less than ρ	Name
0	0.0000	–
0.6745	0.5000	Probable error
1.0000	0.6827	Standard error
1.6449	0.9000	Reliable error
1.9600	0.9500	95% error
2.0000	0.9545	–
2.5758	0.9900	–
3.0000	0.9973	–

Thus 68.3% of the time the measured rate will be between 890.0 and 910 cpm. The reliable error (90%) is given by:

$$\frac{1.645}{(8100)^{1/2}} = \frac{1}{90} = 0.0183$$

$$(900)(0.0183) = 16.5 \qquad (900 \pm 16.5).$$

Thus 9 times out of 10 the measured rate will be between 883.5 and 916.5 cpm.

Combination of Errors

It is often desirable to evaluate the error in a function of several counting rates. From the relation $\sigma = K\sqrt{M}$ certain rules are formulated for the manipulation of experimentally derived quantities. Errors can be combined by these expressions only if they are expressed in the same degree of uncertainty: all standard (68.3%) errors or all reliable (90%) errors.

Let a be the error in A and b the error in B: $A \pm a$, $B \pm b$.

The sum or difference is given by:

$$(A \pm a) \pm (B \pm b) = (A \pm B) \pm (a^2 + b^2)^{1/2}.$$

The product is given by:

$$(A \pm a)(B \pm b) = AB \pm AB\left(\frac{a^2}{A^2} + \frac{b^2}{B^2}\right)^{1/2} = AB \pm (a^2 B^2 + b^2 A^2)^{1/2}.$$

The ratio is given by:

$$\frac{(A \pm a)}{(B \pm b)} = \left(\frac{A}{B}\right) \pm \frac{A}{B}\left(\frac{a^2}{A^2} + \frac{b^2}{B^2}\right)^{1/2} = \left(\frac{A}{B}\right) \pm \left(\frac{a^2}{B^2} + \frac{A^2}{B^4}b^2\right)^{1/2}.$$

Since a/A and b/B are the fractional errors in calculating errors of products or ratios, the squares of the fractional errors are added.

For a most common case of obtaining the counting rate of a sample from the observations of the count rate of the background and the sample plus background, let m be the observed total counts in time t, and $R = m/t$ rate. Then

$$R \pm r = \frac{m \pm \epsilon}{t} = \frac{m \pm K_i \sqrt{m}}{t}$$

and

$$\text{fractional error in } m = \frac{K_i \sqrt{m}}{m} = \frac{K_i}{\sqrt{m}}$$

$$\text{fractional error in } R = \frac{K_i \sqrt{m/t}}{m/t} = \frac{K_i}{\sqrt{m}}.$$

Example I. Sample plus background = 6250 counts in 10 min, background alone = 900 counts in 45 min. Then the rate of sample plus background is $R_{s+b} = 6250/10 = 625$ cpm. The standard error is given by:

$$r = \frac{K_i \sqrt{m}}{10} = \frac{1(6250)^{1/2}}{10} = 7.9.$$

Therefore,

$$R_{s+b} \pm r = 625 \pm 7.9 \text{ cpm}.$$

The rate of background is $R_b = 900/45 = 20$ cpm. The standard error is given by:

$$r_b = \frac{K_i \sqrt{m}}{45} = \frac{1(900)^{1/2}}{45} = 0.67.$$

Therefore,

$$R_b \pm r_b = 20 \pm 0.7.$$

The rate of the sample alone is $R_s = R_{s+b} - R_b$. The standard error is given by:

$$r_s = \pm(r_{s+b}^2 + r_b^2)^{1/2}.$$

Therefore,

$$R_s \pm r_s = (625 - 20) \pm [(7.9)^2 + (0.7)^2]^{1/2}$$

$$= 605 \pm (63)^{1/2}$$

$$= 605 \pm 7.9.$$

Example II. A sample is counted for 10 min giving 340 cpm. After a chemical manipulation, a 10-min counting period gives 300 cpm. The total number of counts taken was 6400. Then,

$$\sigma = \frac{K(6400)^{1/2}}{20} = K\frac{80}{20} = 4K, \qquad \sigma = \sqrt{m}$$

$$\sigma_R = \frac{\sqrt{m}}{t}, \qquad \sigma_{\text{diff}} = (a^2 + b^2)^{1/2} = \left(\frac{m_1}{t^2} + \frac{m_2}{t^2}\right)^{1/2} = \frac{\sqrt{M}}{t}.$$

The difference is $340 - 300 = 40$ cpm. So

Reliable error: $4 \times 1.645 = 6.6$ cpm, or 40 ± 6.6 cpm difference.
Probable error: $4 \times 0.675 = 2.7$ cpm, or 40 ± 2.7 cpm difference.

Nine chances out of 10 the difference will be between 33.4 and 46.6 cpm. There is a 50–50 chance that the difference will be between 37.3 and 42.7 cpm.

Example III. Two samples, 800 and 1000 cpm. respectively, are counted 10 min each. What is the ratio of the counting rates and its error?

$$A \pm a = 8000 \pm K(8000)^{1/2}, \qquad B \pm b = 10{,}000 \pm K(10{,}000)^{1/2},$$

$$\left(\frac{A}{B}\right) = \left(\frac{A}{B}\right) \pm \left(\frac{A}{B}\right)\left(\frac{K^2A}{A^2} + \frac{K^2B}{B^2}\right)^{1/2} = \left(\frac{A}{B}\right) = K\left(\frac{A}{B}\right)\left(\frac{1}{A} + \frac{1}{B}\right)^{1/2}$$

$$= \frac{8000}{10{,}000} \pm K(0.8)\left(\frac{1}{8000} + \frac{1}{10{,}000}\right)^{1/2}$$

$$= 0.8 \pm 0.8K[(1.25 + 1.00) \times 10^{-4}]^{1/2}$$

$$= 0.8 \pm 0.8K(1.5 \times 10^{-2})$$

$$= 0.8 \pm 0.012K.$$

Reliable error: $K = 1.645$,   or ratio: 0.8 ± 0.0197.
Standard error: $K = 1.000$,   or ratio: 0.8 ± 0.012.
Probable error: $K = 0.675$,   or ratio: 0.8 ± 0.0081.

Time Requirements

Percentage error in a counting rate, E, is given by

$$\sigma_R = \frac{K\sqrt{N}}{t} = K\left(\frac{R}{t}\right)^{1/2}.$$

When the known activity is counted for a time t,

$$E = \frac{100K\sqrt{N}}{N} = \frac{100K}{(Rt)^{1/2}}.$$

Therefore the counting time necessary to get a certain percentage error is

$$t = \frac{10^4 K^2}{R(E)^2}$$

Example. How long must an activity of approximately 120 cpm be counted so that it will be known to: (a) a standard error of 1.5%; (b) a reliable error of 1.5%?

$$\text{(a)} \quad t = \frac{10^4(1.00)^2}{(120)(1.5)^2} = 37 \text{ min,}$$

$$\text{(b)} \quad t = \frac{10^4(1.645)^2}{(120)(1.5)^2} = 100 \text{ min.}$$

Dividing Time between Sample and Background

The error due to the presence of the background can be minimized by dividing properly the total counting time between the counts for background and those for sample plus background:

$$\sigma = \left(\frac{R_{s+b}}{t_{s+b}} + \frac{R_b}{t_b}\right)^{1/2}, \qquad \sigma^2 = \frac{R_{s+b}}{t_{s+b}} + \frac{R_b}{t_b}, \qquad \sigma_R = \frac{\sqrt{M}}{t} = \left(\frac{R}{t}\right)^{1/2}.$$

To minimize the error,

$$d\sigma = 0, \qquad d(t_{s+b} + t_b) = 0.$$

To minimize the total counting time,

$$dt_{s+b} = -dt_b, \qquad 2\sigma\, d\sigma = -\frac{R_{s+b}}{t_{s+b}^2}\, dt_{s+b} - \frac{R_b}{t_b^2}\, dt_b = 0,$$

$$\frac{R_{s+b}}{t_{s+b}^2} = \frac{R_b}{t_b^2}, \qquad \text{or} \qquad \frac{t_b}{t_{s+b}} = \left(\frac{R_b}{R_{s+b}}\right)^{1/2}.$$

Example. Suppose sample plus background count approximately 1000 cpm and the background alone counts approximately 2.5 cpm. How should the time available for counting be divided?

$$\frac{t_b}{t_{s+b}} = \left(\frac{2.5}{1000}\right)^{1/2} = \frac{1}{20}.$$

The background need be counted only one-twentieth of the time that the sample is counted.

Division of Counting Time for Desired Error

If one knows the background counting rate, then, for different counting rate samples, the percentage error in the sample counting rate is

$$E_{Rs} = \frac{100K(R_{s+b}/t_{s+b} + R_b/t_b)^{1/2}}{R_{s+b} - R_b}$$

where $R_s = R_{s+b} - R_b$. Also, letting $R_{s+b}/R_b = \gamma$, $\sqrt{\gamma} = t_{s+b}/t_b$,

$$(E_{Rs})^2(R_{s+b} - R_b)^2 = 10^4 K^2 \left(\frac{R_{s+b}t_b + R_b t_{s+b}}{t_{s+b}t_b}\right)$$

$$t_b = \frac{10^4 K^2[(R_{s+b}t_b + R_b t_{s+b})/t_{s+b}]}{(E_{Rs})^2(R_{s+b} - R_b)^2}.$$

Since

$$(R_{s+b} - R_b)^2 = \left(\frac{R_{s+b} - R_b}{R_b}\right)^2 R_b^2 = (\gamma - 1)^2 R_b^2$$

and

$$\left(\frac{R_{s+b}t_b + R_b t_{s+b}}{t_{s+b}}\right)\bigg/ R_b = \frac{R_{s+b}}{R_b}\frac{t_b}{t_{s+b}} + 1 = \gamma \cdot \frac{1}{\sqrt{\gamma}} + 1 = \sqrt{\gamma} + 1$$

$$t_b = \frac{10^4 K^2(\sqrt{\gamma} + 1)}{(E_{Rs})^2(\gamma - 1)^2 R_b}.$$

Table XVIII-4 lists the values of t_{s+b} and t_b for various standard errors ($K = 1$) and values of R_{s+b}/R_b, where R_b is 30 cpm.

Table XVIII-4

Counting times for a given standard error of net activity

$$R_b = 30 \text{ cpm}, \qquad R_s = R_{s+b} - R_b$$

			Standard error in R_s (%)[a]					
			1		2		5		10
R_{s+b}/R_b	t_b (min)	t_{s+b} (min)	t_b	t_{s+b}	t_b	t_{s+b}	t_b	t_{s+b}
1.1	68,300	71,700	17,100	18,000	2730	2870	683	717
1.2	17,500	19,100	4360	4780	698	764	175	191
1.4	4550	5380	1140	1350	182	215	46	54
1.7	1570	2050	393	513	63	82	16	21
2.0	804	1140	201	284	32	46	8	11
2.5	383	606	96	152	15	24	4	6
3.0	228	395	57	99	9	16	2	4
3.5	153	286	38	72	6	11	2	3
4.0	111	222	28	56	4	8	1	2
5.0	68	151	17	38	3	6	1	2
6.0	46	113	12	28	2	5	1	1
7.0	34	89	8	22	1	4	1	1
8.0	26	74	7	18	1	3	1	1
9.0	21	62	5	16	1	3	1	1
10.0	17	54	4	14	1	2	1	1

Calculation: $R_{s+b} = \gamma R_b$

$$t_{s+b} = t_b\sqrt{\gamma}$$

$$t_b = \frac{10^4(\sqrt{\gamma} + 1)K^2}{(\gamma - 1)^2 E^2 R_b}; \qquad K = 1$$

[a] Short times are rounded off to the nearest minute greater than zero.

Another way of expressing division of time is as follows: Starting again from

$$\frac{t_b}{t_{s+b}} = \left(\frac{R_b}{R_{s+b}}\right)^{1/2} \quad \text{and} \quad (E_R) = \frac{10^2\sigma_s K}{R_s} = 10^2 K \frac{(R_{s+b}/t_{s+b} + R_b/t_b)^{1/2}}{R_s}$$

$$(E_R)^2 R_s^2 = 10^4 K^2 \left[\frac{R_{s+b}}{t_{s+b}} + \frac{R_b}{t_b}\right] = \left[\frac{R_{s+b}t_b + R_b t_{s+b}}{t_{s+b}t_b}\right] 10^4 K^2$$

or

$$t_{s+b} = \frac{[(R_{s+b}t_b + R_b t_{s+b})/t_b]10^4 K^2}{(E_R)^2 R_s^2} = \frac{[R_{s+b} + R_b(t_{s+b}/t_b)]10^4 K^2}{(E_R)^2 R_s^2}$$

$$= \frac{[R_{s+b} + R_b(R_{s+b}/R_b)^{1/2}]10^4 K^2}{(E_R)^2 R^2} = \frac{(R_{s+b} + \sqrt{R_b}\sqrt{R_{s+b}})10^4 K^2}{(E_R)^2 R_s^2}$$

and

$$t_b = \frac{[(R_{s+b}t_b + R_b t_{s+b})/t_{s+b}]10^4 K^2}{(E_R)^2 R_s^2} = \frac{[R_b + R_{s+b}(t_b/t_{s+b})]10^4 K^2}{(E_R)^2 R_s^2}$$

$$= \frac{[R_b + \sqrt{R_{s+b}}\sqrt{R_b}]10^4 K^2}{(E_R)^2 R_s^2}.$$

For equal predetermined times, $t_{s+b} = t_b$

$$(E_R)^2 R_s^2 = \left[\frac{R_{s+b}}{t_{s+b}} + \frac{R_b}{t_b}\right]10^4 K^2 = \frac{(R_{s+b} + R_b)10^4 K^2}{t}$$

$$t = \frac{(R_{s+b} + R_b)10^4 K^2}{(E_R)^2 R_s^2}$$

Example.   If equal time is needed for background and sample plus background, how long will the count time be to get 2% standard error?

$$R_{s+b} = 300 \text{ cpm}, \qquad R_b = 15 \text{ cpm},$$

$$t = \frac{10^4(1)^2(315)}{(2)^2(285)^2} = \frac{3.15 \times 10^6}{3.25 \times 10^5} = 9.7 \text{ min for each.}$$

For equal predetermined counts, $N_{s+b} = N_b$.   In this case the number of counts for sample plus background equals the number of counts for background alone.

$$(E_R)^2 R_s^2 = \left[\frac{R_{s+b}}{t_{s+b}} + \frac{R_b}{t_b}\right]10^4 K^2 = \left[\frac{R_{s+b}^2}{R_{s+b}t_{s+b}} + \frac{R_b^2}{R_b t_b}\right]10^4 K^2.$$

Since $R_{s+b}t_{s+b} = N_{s+b}$, $R_b t_b = N_b$, $R_b t_b = R_{s+b}t_{s+b}$, then

$$(E_R)^2 R_s^2 = \left[\frac{R_{s+b}^2 + R_b^2}{N_{s+b}}\right]10^4 K^2$$

$$N_{s+b} = \frac{(R_{s+b}^2 + R_b^2)10^4 K^2}{(E_R)^2 R_s^2}.$$

Example. Consider again the case where sample plus background rate was approximately 300 cpm, and background alone was approximately 15 cpm. How many counts, for equal total counts of sample plus background and background alone, are necessary to obtain a 2% standard error?

$$R_{s+b} = 300, \qquad R_b = 15, \qquad R_s = 285;$$

$$N_{s+b} = \frac{[(300)^2 + (15)^2]10^4(1)^2}{(2)^2(285)^2} = \frac{9.02 \times 10^8}{3.30 \times 10^5} = 2.73 \times 10^3.$$

Therefore, it is necessary to count until 2730 counts each of the sample plus background and the background alone have been accumulated.

The Advantage of the 100-min Running Count for Low-Level Counting

Consider a sample which has been counted for 1000 min compared to the same sample which has been counted for 100 min each count for a total of 1000 min. Since the standard deviation σ of a count is the square root of the total number of counts N, i.e., $\sigma = \sqrt{N}$, the fractional error (%) is given by

$$\sigma_f = \frac{\sqrt{N}}{N} \times 100 = \frac{100}{\sqrt{N}}.$$

The 2σ error is just twice the σ value, or in fractional terms, twice σ_f.

The significance of σ (or σ_f) is that 68.3% of all measured values should fall within the error limits. The 2σ error indicates the limits of the error, which will include 95.5% of all the measurements (often referred to as the 95% confidence level).

Example. Consider a counting system which has a background of 22.0 ± 0.1 cpm. A sample was measured which gave a total of 23,000 counts in 1000 min. The same sample was counted for ten 100-min periods with the results shown in Table XVIII-5.

Both counting sequences give the same statistical accuracy to the determination of the counting rate of the sample and background:

$$\sigma_{s+b} = (23,000)^{1/2} = 152 \text{ counts}, \qquad \sigma_f = 0.7\%,$$

and the counting rate is equal to

$$(23,000 \pm 152)/1000 = 23.0 \pm 0.1_5 \text{ cpm}.$$

Table XVIII-5

The results of ten 100-min counts on a sample

Period	Time (min)	Counts	σ
1	100	2287	± 48
2	100	2317	± 48
3	100	2286	± 48
4	100	2295	± 48
5	100	2331	± 48
6	100	2307	± 48
7	100	2306	± 48
8	100	2279	± 48
9	100	2290	± 48
10	100	2302	± 48
Total	1000	23,000	± 152

The sample counting rate is given by the difference from the background rate:

$$(23.0 \pm 22.0) \pm [(\sigma_{s+b})^2 + (\sigma_b)^2]^{1/2},$$

$$1.0 \pm \sqrt{(0.1_5)^2 + (0.1)^2},$$

$$1.0 \pm \sqrt{0.033},$$

$$1.0 \pm 0.2 \text{ cpm } (\sigma \text{ error}),$$

or

$$1.0 \pm 0.4 \text{ cpm } (2\sigma \text{ error}).$$

The important advantage to making a running count involves the problems of spurious events which are not associated with the sample counts or the normal background. These excursions could be the result of electrical factors (power failure, static burst, etc.) or external radiation (cosmic ray burst, someone moving a radioactive source near the counter, etc.). If a single 1000-min count were made, all of the counting time would have been lost; or if the occurrence of the excursion were unknown, an incorrect sample count rate would be reported. If the excursion which occurred within one of the 100-min segments were large enough to give a count rate significantly different from the count rate for the other 100-min segments, the data for that segment could be eliminated without rejection of the other nine segments of data. That means that nine-tenths of the data was saved. An example is given in Table XVIII-6.

Table XVIII-6

The results of ten 100-min counts on a sample with one datum which appears to be outside error limits[a]

Period	Time (min)	Counts	σ	Difference from average
1	100	2310	± 48	-90
2	100	3280	± 57	$+880$
3	100	2275	± 48	-125
4	100	2298	± 48	-102
5	100	2307	± 48	-93
6	100	2315	± 48	-85
7	100	2323	± 48	-77
8	100	2287	± 48	-113
9	100	2291	± 48	-109
10	100	2316	± 48	-84
Total	1000	24,002	± 155	

[a] Background $= 22.0 \pm 0.1$ cpm.

There are many methods of testing a set of data to determine the reliability of each datum. Let us just accept the fact that the data for period 2 are outside the normal statistical variation expected and can be eliminated from consideration for calculation of the sample activity.

If a single 1000-min counting period had been recorded, the sample activity would be calculated as:

$$(24.0 - 22.0) \pm [(0.1_6)^2 + (0.1)^2]^{1/2} \qquad \text{or} \qquad 2.0 \pm 0.2 \text{ cpm.}$$

For 10 counting periods, without rejection of any of the data, the sample count rate would be calculated as the same, 2.0 ± 0.2 cpm. However, with the rejection of the counts recorded during the second count period as being outside the statistical limits of normal fluctuations due to counting statistics, the sample count rate would be calculated from

$$\text{total counts} = 20{,}722 \pm 144, \qquad \text{time} = 900 \text{ min,}$$

and

$$\text{sample count rate} = (23.0 - 22.0) \pm [(0.1_6)^2 + (0.1)^2]^{1/2} = 1.0 \pm 0.2 \text{ cpm.}$$

Thus the undetected excursion introduced a 100% error in the measured sample count rate which was well outside the statistical probability. The 2σ error would only have been $\pm 40\%$.

This example demonstrates that a repeated 100-min counting period will give more reliable sample counting data than a single counting period for the same total elapsed time. [*Note.* The time lost during printout of the data at the end of each 100-min count is negligible.]

Comparison of Two Assay Systems

One may compare two assay systems of different sensitivities in terms of the total counting time required to obtain a given percentage error in the sample counting rate, E_{Rs}. Let F.M. be a relative figure of merit between the two counting systems such that

$$\text{F.M.} = \frac{(t_{s+b})_1}{(t_{s+b})_2}.$$

A large value of F.M. would mean that system 1 was inferior, on a counting time basis, to system 2.

Using the relationships

$$t_{s+b} = \frac{(R_{s+b} + R_b)10^4 K^2}{(E_{Rs})^2 R_s^2} \quad \text{and} \quad R_{s+b} = R_s + R_b,$$

then

$$t_{s+b} = \frac{(R_s + 2R_b)10^4 K^2}{(E_{Rs})^2 R_s^2}$$

for equal (E_{Rs}) error in both systems expressed in some form (i.e., standard error)

$$\frac{(t_{s+b})_1}{(t_{s+b})_2} = \frac{(R_s + 2R_b)_1 10^4 K^2}{(E_{Rs})_1^2 (R_s^2)_1} \bigg/ \frac{(R_s + 2R_b)_2 10^4 K^2}{(E_{Rs})_2^2 (R_s^2)_2}$$

$$= \frac{(R_s + 2R_b)_1}{(R_s + 2R_b)_2} \frac{(R_s^2)_2}{(R_s^2)_1}.$$

For a "hot" sample where $R_{s+b} \gg R_b$,

$$\text{F.M.} = \frac{(R_s)_2}{(R_s)_1} = \frac{(t_{s+b})_1}{(t_{s+b})_2},$$

which says that the shorter assay time is required for the system having the greater response to the sample.

For a very weak sample where $R_b > R_s$,

$$\text{F.M.} = \frac{(R_s^2)_2}{(R_s^2)_1} \frac{(R_b)_1}{(R_b)_2}.$$

Often in an assay system a change designed to increase R_s will also produce a proportionate increase in R_b. Any system in which R_s can be increased more rapidly than R_b will require shorter counting time for equal errors. Thus gamma counting in a scintillation counter [i.e., NaI(Tl) crystal] may be faster than counting for the same precision in a G–M (Geiger–Müller) tube:

$$(R_s)_{crystal} \gg (R_s)_{G-M}$$

also

$$(R_b)_{crystal} \gg (R_b)_{G-M}$$

but

$$\frac{(R_s)_{crystal}}{(R_s)_{G-M}} \gg \frac{(R_b)_{crystal}}{(R_b)_{G-M}}.$$

Treatment of a Series of Independent Determinations

For a series of independent determinations, $A_i \pm a_i$, the method of averaging by least squares can be used to obtain the average value of A, $\bar{A}$, and its standard deviation, $\sigma(\bar{A})$

$$\sigma^2(A_i) = a_i{}^2 = 1/W_i, \qquad \sigma^2(\bar{A}) = 1/\sum W_i, \qquad \bar{A} = \sum(W_i A_i)/\sum W_i.$$

Example. For the following series of independent determinations, find by the method of averaging by least squares the average and its error (all errors have to be expressed in the same form):

	W_i	$W_i A_i$
4570 ± 200	2.50×10^{-5}	0.114
4430 ± 250	1.74	0.077
4525 ± 170	3.40	0.154
4610 ± 130	5.92	0.273
4560 ± 150	4.45	0.203
	18.01×10^{-5}	0.821

$$\sigma^2(\bar{A}) = \frac{1}{0.1801 \times 10^{-3}} = 5.55 \times 10^3, \qquad \bar{A} = \frac{0.821}{0.1801 \times 10^{-3}} = 4.56 \times 10^3$$

$$\sigma(\bar{A}) = 74.5, \qquad \bar{A} \pm \sigma(\bar{A}) = 4560 \pm 74.5.$$

Data Tests for Reliability

Every piece of data has to be considered as a legitimate measurement, unless the experimenter has some reason for suspecting it. Many variables are associated with activity measurements, and occasionally a result will be obtained which differs greatly from a group of observations as a whole. If this variant result is one of a small number of observations, it may exert an unduly large influence on the average value of the measurement. Often some method is desirable to calculate the probability that you are eliminating a valid piece of data if you reject the variant result.

The most commonly used test is the Chauvenet criterion, although this test has been shown to reject "good" data too often.

There are two other tests that are quite reliable, the outlier test for Poisson data, and Dixon's test for Gaussian and not Poisson data.

Outlier Test.   When the data have a Poisson distribution rejection boundaries can be set such that:

$$X_U = \bar{x} + B_1\sqrt{\bar{x}} \quad \text{and} \quad X_L = \bar{x} - B_1\sqrt{\bar{x}},$$

where X_U and X_L are the upper and lower limits, respectively, and B_1 is the boundary value for a given known error σ.

Example.   Consider the following set of data: 1629, 1605, 1711, 1589, 1567, 1588, 1598, 1569, 1651, 1556, 1576, 1611, 1606, 1635, 1613, 1627:

$$\bar{x} = 1608.2, \quad \sqrt{\bar{x}} = 40.1.$$

To test the outlier value 1711

$$u = \frac{1711 - 1608.2}{40.1} = 2.564$$

from the table of B_1 values for probability α (Table XVIII-7) and a number of observations

$$\text{at } \alpha = 0.05: \quad B_1 = 2.64, \quad u < B_1, \text{ not rejected,}$$

$$\text{at } \alpha = 0.10: \quad B_1 = 2.40, \quad u > B_1, \text{ rejected.}$$

The actual $\alpha = 0.065$, which means that 6.5% of time you are rejecting a good datum. This is a one-sided test where only high or low variations are

Table XVIII-7

Table of boundary values B_1 for outlier test with known α

	α				
n	0.10	0.05	0.025	0.01	0.005
3	1.50	1.74	1.95	2.22	2.40
4	1.70	1.94	2.16	2.43	2.62
6	1.94	2.18	2.41	2.68	2.87
8	2.09	2.33	2.56	2.83	3.02
10	2.20	2.44	2.66	2.93	3.12
15	2.38	2.62	2.84	3.10	3.29
20	2.50	2.73	2.95	3.21	3.39
25	2.59	2.82	3.03	3.28	3.47

expected. For two-sided tests you have to replace α with 2α, namely, you can reject 1711 only at the 13% level if it is possible to get both high and low values. At this level the value should not be rejected, since 13% of time you are rejecting good data.

one-sided test—93.5% outside,   6.5% inside the limits

two-sided test—87% outside,   13% inside the limits.

Rejection of 1711 is advisable for a one-sided experimental result, leaving a new average

$$\bar{x}_r = 1601.3.$$

The shift in average is comparable to the counting error

$$\bar{x} - \bar{x}_r = 1608.2 - 1601.3 = 6.9$$

and

$$\sigma(\bar{x}_r) = \frac{(\sum x_i)^{1/2}}{15} = \frac{(24020)^{1/2}}{15} = 10.3.$$

Dixon's r Ratios. This method depends on the ratio of the difference between adjacent or almost adjacent end values and the range of values. All data are arranged in increasing order of their value, where x_1 is the lowest value, x_2 the next highest, and so on to the highest value x_n. See Table XVIII-8.

Table XVIII-8

Table of R_{ij} values for Dixon's outlier test

			α				
n	0.20	0.10	0.05	0.02	0.01	0.005	Criterion
3	0.78	0.89	0.94	0.98	0.99	0.99	
4	0.56	0.68	0.77	0.85	0.89	0.93	r_{10}
5	0.45	0.56	0.64	0.73	0.78	0.82	
7	0.34	0.43	0.51	0.59	0.64	0.68	
8	0.39	0.48	0.55	0.63	0.68	0.73	r_{11}
10	0.33	0.41	0.48	0.55	0.60	0.64	
11	0.44	0.52	0.58	0.64	0.68	0.71	r_{21}
13	0.40	0.47	0.52	0.58	0.62	0.65	
14	0.42	0.49	0.55	0.60	0.64	0.67	
16	0.39	0.45	0.51	0.56	0.60	0.62	
18	0.36	0.42	0.48	0.53	0.56	0.59	r_{22}
20	0.34	0.40	0.45	0.50	0.54	0.56	
25	0.30	0.36	0.41	0.46	0.49	0.52	

Consider the aforementioned data:

n: 1,   2,   3,   4,   5,   6,   7,   8,
datum: 1556, 1567, 1569, 1576, 1588, 1589, 1598, 1605,

n: 9,   10,   11,   12,   13,   14,   15,   16
datum: 1606, 1611, 1613, 1627, 1629, 1635, 1651, 1711.

The Dixon outlier r criteria are as follows:

$$r_{10} = \frac{x_2 - x_1}{x_n - x_1} \quad \text{or} \quad \frac{x_n - x_{n-1}}{x_n - x_1}, \qquad r_{20} = \frac{x_3 - x_1}{x_n - x_1} \quad \text{or} \quad \frac{x_n - x_{n-2}}{x_n - x_1},$$

$$r_{11} = \frac{x_2 - x_1}{x_{n-1} - x_1} \quad \text{or} \quad \frac{x_n - x_{n-1}}{x_n - x_2}, \qquad r_{21} = \frac{x_3 - x_1}{x_{n-1} - x_1} \quad \text{or} \quad \frac{x_n - x_{n-2}}{x_n - x_2},$$

$$r_{12} = \frac{x_2 - x_1}{x_{n-2} - x_1} \quad \text{or} \quad \frac{x_n - x_{n-1}}{x_n - x_3}, \qquad r_{22} = \frac{x_3 - x_1}{x_{n-2} - x_1} \quad \text{or} \quad \frac{x_n - x_{n-2}}{x_n - x_3}.$$

Again consider the datum 1711. Calculate r_{22}:

$$r_{22} = \frac{1711 - 1635}{1711 - 1569} = 0.535,$$

$$\text{at } \alpha = 0.05: \quad r_{22} = 0.51,$$

$$\text{at } \alpha = 0.02: \quad r_{22} = 0.56.$$

Therefore, between 2 and 5% of the time elimination of datum 1711 will reject a good datum; less than 5% of the time it will reject a good datum. If it is possible to have high and low values, 2α has to be substituted for α and then about 10% of the time a good datum will be rejected.

AUTHOR INDEX

Numbers in parentheses are reference numbers and indicate that an author's work is referred to although his name is not cited in the text. Numbers in italics show the page on which the complete reference is listed.

A

Ackerman, M. E., 63(35), *68*
Ageno, M., 2(4), *10*
Alexander, T. K., 280(18), *289*
Allred, J. B., 159(89), *173*
Alvarez, J., 153(17), 176(10), *171, 187*
Anastassiadis, P. A., 156(63), *173*
Anbar, M., 155(42), *172*
Anderson, E. C., 300(3), 302(3), 303(3), 304 (3, 6, 7), 305(3), *305*
Anderson, C. E., 119(33), *144*, 282(22), 284 (22), *289*
Anliker, R., 57(33), 63(33), *68*
Armstrong, F. E., 210(4), *225*
Arnold, J. R., 3, *10*, 223, *226*
Ashoroft, J., 117(30), 118(30), *144*
Avinur, P., 155(40, 41), *172*
Axtman, R. C., 284(26), *289*

B

Baden, H. P., 176(15), *187*
Baillie, L. A., 74(6), *89*, 211(5), *225*
Barnes, R. F., 123(40), *144*
Basson, J. K., 16(2), *33*, 100(15), 102(16), *143*
Batchelor, R., 92(4), 119(32), *143, 144*
Baukema, J., 176(18), *187*
Baxter, C. F., 156(58), *173*, 176(16), *187*
Beard, G. B., 125(42), *144*
Belcher, E. H., 153(19), *171*
Belikova, T. P., 6, *10*
Benson, R. H., 160(94), *173*
Berlin, N. J., 153(24), 158(24), *172*
Berlman, I. B., 21(24), 28(32), *33*, 50(17), 52(22), 64(38), *67, 68*, 104(22), *143*
Bernal, L., 153(27), *172*
Berson, S. A., 258(1), *262*
Bibrom, R., 154(34, 35), *172*

333

SUBJECT INDEX

A

Absorption spectrum, electronic bands, 18
Acid solubilization, 153–154
Acrylamide gels, 158
Alpha–beta ratio, 104–105
Alpha emitters, 100–111
Alpha particle,
  counting efficiency, 102
  energy resolution, 105–109
  linewidth, 105–106
  low-level counting, 110–111
  relative response of, 92–94
Amplifiers,
  linear, 69–71
  logarithmic, 69–71
Anode, 85–88
  fatigue, 86–88
  gain recovery, 87–88
  saturation, 86–88

Aromatic solvents,
  excited-state yields, 43
Auger electrons, 98–100

B

Background, 198–207
  Cerenkov radiation, 201–202
  chemiluminescence, 204–206
  cosmic radiation, 201–202
  cross talk, 200–201
  natural radioactivity, 202–204
  natural tritium and carbon-14, 203–204
  noise, 198–200
  photoluminescence, 206–207
  sources, 199
Balance-point counting, 223
BBOT, fluorescence properties, 48–49
Benzene, relative response, 37–38

A 4
B 5
C 6
D 7
E 8
F 9
G 0
H 1
I 2
J 3